Imitation of Rigor

Imitation of Rigor

An Alternative History of Analytic Philosophy

MARK WILSON

OXFORD
UNIVERSITY PRESS

Great Clarendon Street, Oxford, OX2 6DP,
United Kingdom

Oxford University Press is a department of the University of Oxford.
It furthers the University's objective of excellence in research, scholarship,
and education by publishing worldwide. Oxford is a registered trade mark of
Oxford University Press in the UK and in certain other countries

First Edition published in 2021

Impression: 1

Published in the United States of America by Oxford University Press
198 Madison Avenue, New York, NY 10016, United States of America

British Library Cataloguing in Publication Data

Data available

Library of Congress Control Number: 2021943845

ISBN 978-0-19-289646-9

DOI: 10.1093/oso/9780192896469.001.0001

Printed and bound in the UK by
TJ Books Limited

To my brother George

Contents

Synopsis

Chapter 1: Ersatz Rigor

Our discussion charts a methodological path that philosophy might have followed, if it hadn't become sidetracked by the faulty promises of a misconceived standard of "rigor." To diagnose what has gone wrong, we will first revisit a critical, but nearly forgotten, episode within the prehistory of analytic philosophy. It concerns a seemingly insignificant puzzle connected with the notion of "force" that blocked the advance of physical knowledge in the late nineteenth century. Perceptive philosopher/scientists of the time (notably Heinrich Hertz and Ernst Mach) decided that conceptual clashes of this general character represent the inevitable outcome of any descriptive practice that enlarges itself through the gradual accretion of extended reasoning techniques. Often this increasing complexity becomes ingrained within a linguistic usage without attracting the overt attention of the speakers in question. When it eventually does, questions of the form "what does a usage of this character actually tell us about the external world?" often arise. The book categorizes such queries as "small metaphysics concerns" because they directly ask how successful descriptive practices manage to encode sound data with respect to objective reality in helpful ways. Traditionally such questions have been either rejected as "meaningless" by anti-metaphysicians or supplied with simple "Fido"/Fido answers by "realists" (but one of this book's projects is to introduce its readers to more sophisticated modes of data registration in Chapter 5, as a helpful expansion of "small metaphysics" possibility). Historically, the familiar systems of grander metaphysics often sprouted from these modest seeds within mathematics and the philosophy of language. In doing so, these inflated schemes frequently overshot their targets and imposed unacceptable methodological restrictions upon the free development of science.

These were the methodological circumstances that Hertz and Mach confronted. They needed to somehow ameliorate the "small metaphysics" obstacles that they encountered within mechanical practice without endorsing the strictures of any overbearing philosophical scheme. In his 1894 *Principles of Mechanics*, Hertz suggested that the impasse associated with "force" could be bridged if classical mechanics could carve back the collection of doctrines it associates with the word until a suitably pruned sub-selection assumes the self-consistent format of an axiomatized theory. Through such reconstructions, skilled methodologists could surmount the traditional inconsistences encountered within our evolved descriptive practices without any dogmatic reliance upon "grand metaphysics" themes that might compromise a scientist's ability to theorize freely.

Hertz's methodological proposals struck subsequent generations as compelling and the presumption that sound science should always aim for the crisp contours of an axiomatized theory became deeply ingrained within the philosophical psyche (the book labels these propensities as "theory T thinking"). Recasting a confusing practice into the disciplined confines of a logically ordered "theory T" thus became widely accepted as the central manner in which a philosopher could strive for precision and clarity. But this book will argue that this circumscribed conception of "rigor" is ill-suited to the conceptual puzzles that troubled Hertz in the first place.

In the 1930's, Rudolf Carnap and his associates went further and claimed that Hertz's axiomatized conception of a self-contained linguistic system could liberate science (and philosophy itself) from any concern with "metaphysics" whatsoever, great or small. In many ways, this position extended the distrust that Hertz and Mach had articulated with respect to the undesirable imperatives of traditional dogmatic metaphysics. In more recent times, self-styled "analytic metaphysicians" have attempted to reinstate many of the grander methodological doctrines of earlier times by repackaging them as "theories" of a suitably rigorous character. We will argue that these revised doctrines underestimate the subtleties of free scientific inquiry in the same manner that Hertz and Mach opposed. But the smaller bore "philosophy of language and mathematics" questions at their root demand non-trivial answers.

To unravel these tangled threads, we must distinguish the natural developmental factors that Hertz and Mach *diagnose* as responsible for the problems with "force" from the axiomatic *remedies* that Hertz proposes for their amelioration. In Chapter 5, we will find that modern work in computer modeling suggests a more sophisticated "architectural" resolution to Hertz's difficulties with "force." It constitutes a vein of diagnostic inquiry that suggests a middle path that philosophical investigation might follow that avoids the dogmatisms characteristic of both sides of our current metaphysics-versus-anti-metaphysics divisions. It is this same "middle path," rooted in Mach's and Hertz's original ruminations upon the puzzles of "force," that suggests the counterfactual developments that analytic philosophy might have preferably followed.

Chapter 2: Prospectus

The methodological lessons of this book are broadly intended, but they can be effectively illustrated only through the consideration of concrete problems. To this purpose, we will first revisit the specific "small metaphysics" puzzles that troubled Hertz. In doing so, we can directly benefit from his profound insights into the linguistic processes that generate these difficulties. As noted previously, Hertz and Mach regard the natural development of language as evolutionary in character, acquiring fresh capacities and skills along the way without great concern for conceptual unity. The eventual result is frequently a descriptive practice that encodes physical data in effective but deviously complex ways. This diagnosis of the origins of conceptual confusion strikes me as essentially correct, but the axiomatic correctives that Hertz endorses do not. In Chapter 5, we'll find that more sophisticated forms of linguistic framework can retain all of the practical advantages of the original practice while nonetheless subduing the internal conflicts that generate the problematic tensions.

Hertz and Mach approach the problems of linguistic usage in a strongly pragmatist spirit, and this book hews to these same methodological guidelines as well.

Chapter 3: Inductive Warrant

Before we consider Hertz's diagnostic insights in detail, we must first review why his underlying concerns seemed critical to the advance of classical mechanics in the late nineteenth century. Unfortunately since that time a pervasive presumption that "classical mechanics is simple" has constricted the intellectual horizons of most observers who no longer actively participate in its continuing applications to practical problems. As a result, Hertz's motivating concerns disappear, not because they have been resolved, but simply because they have been ignored. This collective amnesia leads many contemporary philosophers to confidently assume that they understand what "the worlds of classical mechanics are like," although these rash presumptions embody a significant degree of simplistic misrepresentation. Accordingly, the present chapter outlines the forgotten background required to appreciate Hertz's intellectual puzzles as he confronted them. These details are not required for the central argument of the book, but they nicely illustrate the natural contexts from which "small metaphysics" puzzles characteristically emerge within a gradually evolving discourse.

Chapter 4: The Mystery of Physics 101

As we have already outlined, Hertz and his contemporaries viewed conceptual disharmony as the inevitable product of the evolutionary manner in which an initial descriptive practice gradually extends its applicational outreach, pragmatically guided by the acquisition of fresh methods for calculating results in a useful manner. As a side effect of this gradual accumulation of technique, component words become naturally pulled into subtly different forms of localized referential attachment. These discordant associations create difficulties whenever a straightforward exposition of "fundamental principle" is required, as arises when we attempt to teach an elementary class in classical mechanics (this is the "mystery" of the chapter's title). However, the finer details of Hertz's closely observed discussion have been frequently misunderstood by later commentators, so the bulk of this chapter is devoted to rectifying these issues. Once this is accomplished, the applicability of the multiscalar resolution suggested in the succeeding chapter becomes quite immediate. Again the finer details canvassed in this chapter are not essential for the main argument of the book, but they nicely illuminate its motivational background.

Chapter 5: Multiscalar Architectures

Many of the recent advances in computing capacity have been facilitated by the development of modeling architectures that implement a crucial division in descriptive labor by asking various varieties of submodeling to work together in cooperative harmony without engaging in a straightforward amalgamation of their respective conclusions. To this purpose, the submodels are segregated by characteristic scale length so that a submodeling that focuses exclusively upon behaviors that normally emerge upon a macroscopic length scale (Δ_H) will find its predictions mollified later by corrective messages that originate within a submodel that concentrates upon the local events that arise upon a shorter scale length (Δ_L). Typically, these submodels employ the same descriptive vocabulary (e.g., "force") but do so in variant manners that can generate problematic inconsistences if their specialized origins are not attended to. Multiscalar schemes avoid these clashes by keeping their submodels carefully segregated from one another and only allowing specialized corrective messages (called "homogenizations") to pass between them. The complete multiscalar scheme then requires that its submodels reach an eventual descriptive accord with respect to the corrective messages that they exchange amongst themselves, but never asks them to agree upon the localized claims they advance with respect to "force" per se. The efficiencies obtained through this simple expedient are truly astonishing and have greatly expanded science's capacities for modeling complex systems in effective ways. If so, shouldn't we anticipate that the descriptive policies we follow in our everyday linguistic practices will avail themselves of similar strategic advantages, even if we fail to notice the architectural arrangements required to maintain the overall coherence of such discourses? Such considerations point to a wider range of potential applications for our multiscalar constructions, although we shall largely confine our attention to scientific exemplars in this book.

In any case, we can immediately apply these insights to Hertz's original dilemmas with respect to "force" and find that standard classical practice likewise utilizes submodels that tacitly employ that word in scale-sensitive ways. These unnoticed reasoning arrangements then supply a better resolution of our "Mystery of Physics 101" tensions. Rigorization through straightforward theory T regimentation fails in such circumstances, because the essential key to multiscalar success lies in not attempting to amalgamate all of its localized conclusions into a single-leveled discourse suitable for axiomatization.

A multiscalar architecture can be quite precisely delineated, but it will not exemplify the unsuitable form of syntactic "rigor" that Theory T thinking demands. By elevating the latter to preeminent status as the required goal of analytic philosophy, the subject deprives itself of the very tools required to unravel the inevitable puzzles of real life discourse in a credible fashion.

Chapter 6: Diversity in "Cause"

This chapter extends the diagnostic lessons of the previous chapter to various contemporary controversies with respect to causation, in which the word "cause" appears to highlight different forms of physical consideration depending upon the context in which it is employed. By carefully reviewing a standard treatment of billiard ball behavior from a multiscalar point of view, it becomes easy to appreciate the reasons why "cause" naturally adjusts its referential significance according to context. Attempts to isolate a single underlying significance for "cause" are bound to fail, for the same "division of linguistic labor" reasons that lead the word "force" to its varied forms of referential attachment. In particular, recent attempts to isolate the central "metaphysical" prerequisites of causation are likely to lead to the same ill-advised strictures upon scientific thinking that Hertz and Mach opposed. Instead, we should attend more closely to the "small metaphysics" puzzles that originally generate these grander concerns, viz., questions of the type "what does a usage of this character actually tell us about the external world?" Here we have much to learn, for multiscalar architectures utilize cleverer forms of data encoding than we are likely to have anticipated while lounging in our exclusively "metaphysical" armchairs.

Chapter 7: Dreams of a Final Theory T

Generalizing upon these last remarks, the grander metaphysical schemes popular in Hertz's era frequently suppressed conceptual innovation in manifestly unhelpful ways. In counterreaction, Hertz and his colleagues stressed the raw pragmatic advantages of "good theory" considered as a functional whole and rejected the armchair meditations upon central concepts characteristic of the metaphysical enterprises they spurned. As we also noted, Rudolf Carnap's later rejection of all forms of "metaphysics" attempted to broaden their methodological liberties to a wider canvas. In doing so, the notion of an integrated, axiomatizable "theory" became a central tenet within analytic philosophy's conception of how the enterprise of "conceptual analysis" should be "rigorously" prosecuted. Although Carnap hoped to suppress all forms of metaphysics, large and small, through these means, in more recent times, other veins of "theory T thinking" have instead encouraged a revival of grand metaphysical speculation that embodies many of the suppressive doctrines that Hertz's generation rightly resisted (e.g., the school of "analytic metaphysics" founded by David Lewis). But this opposing side of the metaphysics/anti-metaphysics coin suffers from the same devotion to an inappropriate standard of methodological "rigor." A better approach will pursue the "middle path" suggested by the more accommodating arrangements of a multiscalar architecture.

Chapter 8: Linguistic Scaffolding and Scientific Realism

These studies suggest that philosophical analysis should investigate linguistic adaptation in the same naturalistic spirit as a biologist investigates the environmental adaptations of a particular animal or plant. Doing so effectively requires attending to shaping considerations that arise from a wide variety of strategic sources, including the issues of computational complexity that supply multiscalar tactics with their great advantages. For various reasons that ultimately trace to the theory T thinking that this book rejects, the philosophical notion of "scientific realism" has become unhappily aligned with various simplistic assumptions with respect to the effective forms of word/world alignment. Our conceptions of scientific and linguistic possibility have suffered greatly as a result.

Chapter 9: Truth in a Multiscalar Landscape

Closely aligned with these last considerations are a variety of mistaken presumptions with respect to the evaluative utilities of the word "true's in our linguistic lives. These misconceptions likewise stem from a failure to recognize the tacit architectures that commonly underpin effective discourse. These affinities suggest an intriguing linkage between "true's practical utilities and the "homogenization message" evaluations that are central to the implementation of a multiscalar architecture, for the latter requirements can be naturally captured in scale-length sensitive locutions of the form "If condition S becomes true upon the lower scale length ΔL, then condition S^* will become true on the higher scale length ΔH." Overlooking these non-trivial relationships in the single-leveled manner characteristic of theory T thinking encourages the popular "deflationist" doctrine that maintains that "S is true" conveys no new information beyond what "S" itself says. But the corrective feedback loops characteristic of our richer architectures assign locutions like "S is true" the more substantive task of implementing the revisionary "homogenization messages" that are central to the operations of a multiscalar scheme.

Foreword

This monograph attempts to revive a pragmatically inflected approach to conceptual analysis that was prominent in the early days of what eventually became "analytic philosophy," but which later fell by the wayside. This diminishment is largely the byproduct of the twentieth century's advances in mathematical logic which have supplied an admirably crisp account of what the notion of "rigorously follows from axioms" demands. But as often occurs with any great discovery, its applications were stretched beyond their legitimate reach, leading to a narrowed conception of "philosophical rigor" that is ill-suited to the requirements of effective conceptual remedy. This book will develop this theme by revisiting the trajectory of one of those early axiomatic proposals (Heinrich Hertz's critique of "Newtonian force") and comparing his suggested repairs with the developments that have subsequently emerged within modern computer modeling. This exercise will supply a revealing illustration of the delicate diagnostic work required in unraveling a characteristic philosophical knot. Our concluding chapters will extend these lessons to several wider topics of contemporary metaphysical interest.

Although I had originally planned to include a larger survey of contemporary philosophical writing in this book, I eventually decided that such a critique would only prove tedious and futile, largely because the latent methodological presumptions I shall criticize are rarely enunciated explicitly. Better, I have decided, to return to the original roots of these attitudes within the admirable writings of the philosopher/scientists of the late nineteenth century, who frequently articulated my central concerns in an attractive manner, in tight alignment with the practical roadblocks they were forced to confront within their scientific deliberations. By reconnecting the "philosophical" with the "practical" in this direct manner, my readers should find it easy to develop their own criticisms of the contemporary methodologies that are efficiently outlined in surveys such as Daly (2010) or Williamson (2008).

Although this book largely attends to a specific thread within the early developmental history of "analytic philosophy" so-called, our discussion will highlight vital aspects of that original heritage that deserve revitalization along a wider front (none of our central concerns make any significant appearance within Michael Potter's otherwise excellent volume (2020)). So while this book is not intended as a dedicated history of the era, I hope that it pays grateful tribute to some neglected veins of fertile methodological reflection that characterized those first stirrings of "analytic" endeavor. Indeed, I have advertised its contents as "an alternative history of analytic philosophy" because they illustrate how our subject might have evolved if dubious methodological suppositions hadn't intervened along the way.[1]

In this book I have utilized a number of examples (sometimes the very words!) that have appeared in my other writings (especially Wilson (2017)), and I apologize for the repetitions. But the exposition adopted here is more straightforwardly linear in its narrative, and I hope that its quasi-historical focus will make it easier to follow my general lines of thought. On the other hand, this same devotion to a single motivating example carries its own hazards, because the conceptual landscapes characteristic of "classical mechanical thought" have been severely distorted by latter day commentators who have strived to extract brisk methodological morals from its allegedly simplistic failings. But this mythology of the "illustrative lesson" does not render justice to the complex environments from which the typical conundrums of philosophical concern spring. As a result, a fair portion of Chapters 2 and 3 are devoted to clearing away some of this accumulated rubbish. I feel that an adequate coverage of my favored example demands that we plow through some of these popular misunderstandings, but I also worry that these same details may dissuade the non-specialist reader from reaching the central behaviors that this book attempts to document, viz., the hidden tactics whereby an evolving descriptive practice manages to rectify the inherited tensions that would otherwise undermine its practical utilities. My central exemplar of an insignificant-appearing "reasoning tension" will finally emerge in Chapter 4 ("The Mystery of Physics 101") and the compensating "hidden tactics" are introduced in Chapter 5 ("Multiscalar Architectures"). I believe that this particular puzzle is evocatively illustrative of the general factors that generate allied difficulties everywhere across the broad spectrum of human thought. At present, however, the "wet blanket of philosophical rigorists" (Heaviside 1912, III 370) has greatly discouraged due attention to the conceptual behaviors of this type, so I am hoping that a particularly clean instance of the root phenomena drawn from the real-life career of classical physics may convince them of their oversight.

I am grateful to Gordon Belot, Laura Ruetsche, Tom Ryckman, Sheldon Smith, Alexis Wellwood and Porter Williams for sundry symposium comments and to Bob Batterman, Kathleen Cook, Josh Eisenthal, Anil Gupta, Jennifer Jhun, Juliette Kennedy, Travis McKenna, Bob Pippin, and Jim Woodward for further helpful advice. Penelope Maddy and a group of her students (Adam Chin, Charles Leitz, Chris Mitsch, Jeffrey Schatz, and Evan Sommers) read an earlier version of the manuscript at Irvine, an assistance which has proven very helpful here. A substantial email correspondence with Pen has helped me sharpen my themes further. I remain grateful to my accommodating editor, Peter Momtchiloff.

---

[1] This is also the reason why several otherwise extraneous appendices have been included to correct the historical record.

1

Ersatz Rigor

Logic has nothing to do with it, either with the fact, the discovery, or its use. At the very same time it must be said that a sufficiently profound study of the subject would ultimately lead to the logic of its laws, as a final result. What I do strongly object to is the idea that the logic should come first, or else you prove nothing. Yet perhaps the majority of academical mathematical books are written under this idea. In reality the logic is the very last thing, and that is not final.

Oliver Heaviside (Heaviside 1912, III 370)

Even metaphysicians start small. Or, at least, the best of them have often done so. Some puzzling disharmony arises within the development of an otherwise beneficial descriptive practice that cries out for a satisfactory resolution and blocks further advancement until the troublesome anomalies are identified and resolved. These nagging complexities within practical reasoning may prompt a practitioner to wonder, "What sort of external world can explain why it is wise to maintain thesis X or to confidently claim that tenet Z follows from tenet Y?" From these initial musings, philosophy of a wider-ranging character often emerges. Motivational uncertainties of this stripe can arise within virtually any field of human endeavor, but in this book we shall concentrate upon several significant exemplars drawn from the historical development of physical science.

Why do these puzzles arise so frequently? One of the most salient features of our improving intellectual economy lies in the fact that "our pencils often prove wiser than ourselves," as Euler is alleged to have said. By this, he meant that otherwise errant doodling sometimes suggests pathways of reasoning that allow us to capture natural phenomena within our linguistic nets in fashions that we could not have anticipated. Even after we enter these enriched landscapes and avail ourselves of the bounties they offer, their supportive geographies may still baffle us, for we may have fancied that we've reached India when it is actually the Bahamas. When a cogent elucidation of a practice is finally located, its resolution often proves surprising and unexpected, for words, calculations, and the world can entangle themselves in complicated patterns that we would have never noticed save for the raw goad of improving advantage. There is no general or uniform method for anticipating these difficulties, for their sources can prove as varied as nature's wide ranges of dissimilar behaviors.

Imitation of Rigor: An Alternative History of Analytic Philosophy. Mark Wilson, Oxford University Press. © Mark Wilson 2022.
DOI: 10.1093/oso/9780192896469.003.0001

These originating concerns can be properly categorized as "metaphysical" in their own right, in the sense that they frequently embody *puzzles of informational content*, which we might express as "Our evolved patterns of speaking or calculating manifestly reflect genuine features of the world before us, but the exact nature of the correspondences between words and world that underwrite these successes remain elusive." To be sure, the grand systems of speculative metaphysics that we associate with Descartes, Leibniz, Kant, and so forth stretch far beyond these humble seeds, but the present monograph will focus upon the coordinate utilities of keeping their "small metaphysics" origins firmly in mind as well. We should accordingly applaud those curators of intellectual curiosities who have cultivated an ongoing familiarity with a wide spectrum of previous "small metaphysics" puzzles and their eventual resolutions. Encountering an expert who can tentatively suggest "Um, your difficulties seem reminiscent of the obstacles that hindered W within field X" may supply the vital clue that allows another researcher to successfully unravel the mysteries of topic Z. Training in philosophy and its history can significantly assist in these helpful presentiments. Nonetheless, every fresh conceptual mystery is likely to embody its own range of idiosyncratic details as well.

I believe that academic philosophical deliberation proves most useful when it contributes to a free-thinking appreciation of conceptual variety. However, we are all susceptible to the irrepressible human propensity to declare ourselves unchallenged experts within some favored field of choice, despite the fact that such self-estimations are rarely sustainable over an extended period of time. And so philosophy itself can readily fall victim to these unhelpful pretensions, when it forgets the surprising twists and turns characteristic of real-life intellectual quandary. The "ersatz rigor" of which this book complains represents a misguided self-confidence of this self-anointed character, in which a restricted methodology of "formal clarification" currently dissuades philosophical investigators from penetrating as deeply into the "small metaphysics" sources of a problem as it actually demands. Our academic journals are abundantly populated with articles devoted to "formal epistemology," "formal metaphysics," "formal philosophy of science," and so forth, in which the term "formal" generally evinces an exaggerated faith in the standard discriminations offered within contemporary philosophy of science (axiomatics, first-order logic, and a smattering of set theory—I shall be more precise in my cataloguing later on). But this limited palette of diagnostic tools is rarely adequate to the complexities of the "small metaphysics" concerns that naturally arise within the courses of our improving capacities. I am fundamentally a pragmatic adaptationist at heart; our vocabularies progressively tighten their descriptive grip upon the world around us by stumbling across superior strategies for reasoning productively about it. When we afterwards reflect upon the improvements wrought, we are easily bewildered by the informational novelties we confront. Correct identifications of the subterranean tactics in operation frequently demand the supple varieties of detective sleuthing that we find within our most original philosophical thinkers.

For a variety of reasons (whose historical lineage this book will attempt to outline), many contemporary philosophers have fallen into the habit of believing that they should advance "theories" in much the same way that our colleagues in the sciences allegedly do. Here's a typical proclamation in this pseudo-scientific vein, selected almost at random from the contemporary literature (from Andrew Bacon):

> The theory proposed here is strong and settles many contentious questions of metaphysics: it tells us, for instance, that there are no symmetric fundamental relations and that the fundamental properties and relations can exhibit arbitrary patterns of instantiation. Rather than approaching such questions in a piecemeal way, as is representative of these sorts of investigations, I have advanced a single principle that decides these questions all at once. I think this sort of predictive strength should not be held against the theory: debates in these areas of metaphysics are often quite unconstrained, and it is a virtue when a single coherent picture of reality answers the central questions uniformly. The resultant theory should be adjudicated by the usual standards of philosophical theorizing: according to the merits of strength, simplicity, consistency with the sciences, predictive ability, and so forth.   (Bacon 2020, 538–9)

The final sentence endorses the thesis (strongly associated with the writings of David Lewis (Lewis 1983)) that philosophy should imitate the sciences in adjudicating its own proposals by the same standards of "strength, simplicity and predictive ability" that their colleagues in science allegedly adopt. And the author obviously prides himself in the axiomatic "rigor" with which he has pursued these ambitions. That something must have gone dreadfully wrong is suggested by the conclusions he reaches:

> The theory vindicates the thesis that propositions, properties, and relations may be decomposed into their fundamental components via logical operations and ensures that this decomposition is unique *modulo* certain structural manipulations.... We also obtain the principle that the fundamental are simple, in the sense that they cannot be decomposed into other fundamental properties and relations.   (Bacon 2020, 539)

Although the essay is characteristically example-free, the doctrines announced with respect to "fundamentality" reflect a seductive mythology of scientific practice that I'll call "theory T thinking" in the sequel (in dubious tribute to the expository paragraphs that characteristically begin, "Let T and T′ be two theories that...."). In fact, something even worse has transpired, for the author presumes that science is firmly devoted to uncovering a "fundamental theory of everything," in which Nature's favored fundamental properties become uniquely isolated. Leaving aside these dubious (and fuzzy) assumptions about "fundamental theory" (see Chapter 7), the notion that some unique set of primal properties is desired runs contrary to established methodological wisdom in real life science, which emphasizes the wider insights to be obtained from decomposing a familiar subject into unexpected primitives.[1]

What has gone wrong stems from a fetishization of the term "theory," which has gradually developed over time beginning with some relatively innocent proposals of the late nineteenth century. In general usage, the term "theory" scarcely signifies anything stronger than "a useful proposal," but in the eyes of philosophers the word has gradually taken on the unrealistic set of assumptions characteristic of "theory T thinking." An excellent exemplar of the pitfalls of loose thinking in this vein can be found in the phrase "theory of optics," which philosophers invariably presume comprises a tightly articulated set of doctrines. There is a broad range of "useful proposals" that merit the label "theory of optics," some of them extensively mathematized and some not, but none of them offer the breadth of exclusive coverage that philosophers have come to expect of a "theory." This is because the behavior of what we loosely designate as "visible light" involves an extremely complicated mélange of considerations stretching across a wide range of length and time scales in application to an equally extensive array of potential mediums. We are encouraged to presume that "light of a fixed wavelength" can be straightforwardly parsed as "electromagnetic radiation of a specified frequency" and will be subsequently startled when a more advanced discussion declares that this baseline notion of "frequency" is problematic:

> [A]ll optical fields, whether encountered in nature or generated in the laboratory have some random fluctuations associated with them. The monochromatic sources and monochromatic fields discussed in most optics textbooks are not encountered in real life.   (Wolf 2007, xi)

A significant part of the problem stems from the fact that the notion of "initial condition" becomes rather murky in connection with light, a fact that passes completely unnoted within standard theory T thinking (which presumes that the notion requires no explication whatsoever).

It would not benefit my projected audience much if I were to now launch into a fulsome explication of the delicate physical considerations behind these claims.

---

[1] E.g., the deeper insights into geometry provided by Julius Plücker's Euclidean decompositions into so-called "line geometry" (Wilson 2010).

But I am concerned to break the oppressive spell of "philosophers should imitate what the scientists do" that infuses every word within Bacon's essay. Insofar as I can see, similar standards of surface "rigor" based upon a faulty conception of "science" hinder productive investigative effort within many corners of the philosophical world, commonly inducing an overprizing of ersatz mathematization with respect to, e.g., human decision-making. But every good engineer knows that:

> Long calculations that you do not understand, whether by number crunching or Banach space methods, will land you with bridges that fall down. A theory is only as good as your understanding of it.   (Poston and Stewart 1978, 300)[2]

Proving theorems with respect to an improperly analyzed basis frequently supplies a less valuable contribution to intellectual advance than the recasting of a subject's presumptive foundations in an unforeseen but productive manner. Doing otherwise can secretly encapsulate tiny misapprehensions into a framework that can erupt in significant distortions in unanticipated locales afterward. With respect to "light" and "frequency," perceptive physicists and applied mathematicians have repeatedly proven themselves incredibly deft in diagnosing the unexpected mysteries of these seemingly innocuous terms and have often functioned as better students of linguistic development than the philosophers of language themselves. If "imitating what the scientists do" properly embraces "think about things in unexpected and observant ways," the upshot should prompt a leery distrust of "ersatz rigor" projects such as Bacon's, rather than their embrace.

We should bear in mind that congratulating ourselves on the "rigor" of our operative methodology is an unhelpful yet popular means of asserting undeserved authority within a chosen field. A survey of pseudo-scientific literature (e.g., Martin Gardner's delightful volume (Gardner 1952)) supplies a long list of cranks who have sincerely fancied themselves as paragons of scientific endeavor, self-proclaimed geniuses that have "out-Newtoned Newton" in espousing "rigorous new axioms" (e.g., George Francis Gillette's "all things go straight until they bump"). These surface convictions of self-proclaimed "rigor" invariably conceal a blind unwillingness to advance a proposed practice to a proper level of verifiable test. A cousin illustration of this brash self-confidence can be found in the philosopher David Armstrong's *What is a Law of Nature?*, in which he unabashedly asserts that a "metaphysical" philosopher such as himself can usefully discourse

---

[2] When we return to this quotation in Chapter 7, we'll find that we actually understand the rationale of the constructions they criticize better than they think.

about "what science does" without much background in any topic beyond an undergraduate course in first-order logic:

> [Even if philosophers know no actual laws, they] know the forms which state-
> ments of laws take.... It turns out, as a matter of fact, that the sort of fundamen-
> tal investigations which we are undertaking can largely proceed with mere
> schemata of [the sort "It is a law that Fs are Gs"].... Our abstract formulae
> may actually exhibit the heart of many philosophical problems about laws of
> nature, disentangled from confusing empirical detail. To every subject, its appro-
> priate level of abstraction.   (Armstrong 1983, 6)

In real-life application, the term "law of nature" registers physical considerations of vital importance, but their exact characteristics vary from application to application according to context. By mingling these factors together in Armstrong's indiscrim-inate manner, it becomes impossible to parse what he is talking about in concrete terms. He has blithely abandoned the very applicational details required to inves-tigate his chosen topic in a useful manner.

To be fair, few investigations in "formal metaphysics" prove as palpably detached from concrete illustration as Armstrong's, but their presumed policies of "formal analysis" generally pitch their standards of discrimination at an excessively schematic level. There is a simple historical explanation of how these methodological presumptions became so firmly locked in place. They largely date to the manner in which Rudolf Carnap articulated an alleged opposition between "metaphysics" and "anti-metaphysics" in 1932:

> There have been many opponents of metaphysics from the Greek skeptics to
> the empiricists of the 19th century. Criticisms of very diverse kinds have been
> set forth. Many have declared that the doctrine of metaphysics is false, since it
> contradicts our empirical knowledge. Others have believed it to be uncertain,
> on the ground that its problems transcend the limits of human knowledge.
> Many anti-metaphysicians have declared that occupation with metaphysical
> questions is sterile. Whether or not these questions can be answered, it is at any
> rate unnecessary to worry about them; let us devote ourselves entirely to the
> practical tasks which confront active men every day of their lives!
>
> (Carnap 1932, 60)

A bit later, in his *The Logical Syntax of Language*, he continues:

> *Philosophy is to be replaced by the logic of science*—that is to say, by the logical
> analysis of the concepts and sentences of the sciences, for *the logic of science is
> nothing other than the logical syntax of the language of science*.... The fact that no
> attempts have been made to venture still further from the classical forms is

perhaps due to the widely held opinion that any such deviations must be justified—that is, that the new language-form must be proved to be "correct" and to constitute a faithful rendering of "the true logic." To eliminate [an unnecessary loyalty to pre-established "meanings," and] the pseudo-problems and wearisome controversies which arise as a result of it, is one of the chief tasks of this book.   (Carnap 1934, xiv–xv)[3]

This is not to say that subsequent generations of philosophers have universally accepted Carnap's blunt ban on metaphysical speculation, but they have commonly embraced his presumptions on the format in which a satisfactory response to his qualms should be articulated. In this vein, self-styled analytic metaphysicians have proposed axiomatic formulations of metaphysical tenet in exactly the "scientific" format of which Carnap wholeheartedly approves. But these "formal" presumptions lock their proposals into a schematic package that rarely penetrates to the level of diagnostic detail required.

One of the reasons that I informally distinguish between "small" and "large" forms of metaphysical concern in this book is that I largely concur in Carnap's skepticism with respect to the apriorist presumptions of the subject when it is approached upon a grandly schematic scale. But I do not share his doubts with respect to the value of investigating the varied tactics that allow linguistic systems to connect with the external world in profitable ways. These strike me as questions of mathematical strategy and available opportunity that can be investigated in a robustly scientific vein.

In these respects, there is an important back story to Carnap's remarks which we shall explore in the rest of this book. His strongly "anti-metaphysical" stance traces to a methodological heritage rooted in nineteenth-century physics' attempts to conquer a series of "small metaphysics" obstacles that significantly stymied scientific advance within that era. In 1894, Heinrich Hertz prefaced his posthumous *Principles of Mechanics* (Hertz 1894) with a philosophical manifesto that directly influenced David Hilbert, Ludwig Wittgenstein, Ernst Cassirer, and many later figures (including Carnap himself). In this work, Hertz attempts to resolve various genuine confusions with respect to classical mechanics in a formal manner by employing the axiomatic

Hertz

---

[3] Carnap's presumption that even the "choice of logic" employed within the encompassing axiomatics can be implicitly defined through clear "rules" laces through these passages as well, but we shall ignore these radical suggestions here (which were seldom embraced by his logical empiricist successors).

tactics familiar from Euclid. Although Hertz's specific conceptual recommendations were not widely accepted, his general manner of proceeding axiomatically struck most observers as obviously "the right thing to do," even if the particular details of his specific proposals proved unsuitable. Through this methodological presumption Hertz's axiomatized approach became almost universally accepted as the canonical "formal" framework in which any corrective reconstruction of puzzling doctrine should be articulated, with respect to any anointed topic. The coordinate development of first-order logic as a precisified embodiment of "rigor" further cemented these confident methodological expectations in place. Generally left behind were the nagging considerations that had prompted Hertz's excursus into philosophizing in the first place (although Carnap and the later logical empiricists frequently cited Hertz as an inspiration, they rarely display an accurate awareness of what his motivating difficulties actually involved).

But a funny thing happened on the way to the formalism: it never fully materialized! In John Ford's film *The Man Who Shot Liberty Valance*, a cynical newspaper editor famously comments, "This is the West, sir. When the legend becomes fact, print the legend."[4] This "legend" has lingered on in the form of the restricted framework of "formal philosophical analysis," despite the fact that Hertz's methodological gambit did not pan out as expected. In the context of his time, his proposed pattern of resolution would have seemed the only reasonable response, and it is only the unexpected demands of recalcitrant nature that force more sophisticated forms of conceptual resolution upon us.

In recent years, Hertz's diagnostic difficulties have become cast in a revealing light. Continuing research within applied mathematics and computer science has demonstrated that the classical mechanical practices that had worried Hertz had not been registering their descriptive data within the simple "Fido"/Fido data format required for a smooth axiomatization, but secretly utilize a more advanced form of checks-and-balances encoding. Such "multiscalar architectures" facilitate the astonishing inferential efficiencies that we will survey in Chapter 5. Within these novel forms of conceptual housing, the otherwise puzzling behaviors of terms like "force" and "pressure" become greatly clarified.

As a test case, the pocket history of Hertz's proposals neatly illustrates the inadequacies of "formal" approaches to conceptual analysis. Many of the "small metaphysics" motivations that had troubled Hertz remain relevant to the

---

[4] For a philosophical discussion of *Liberty Valance*, see Pippin (2012).

discussions of "law of nature" and "causality" at the core of contemporary investigation into "formal metaphysics." But the analyses currently proposed rarely address the original concerns at a requisite level of idiosyncratic detail. We will be able to pinpoint these diagnostic inadequacies more clearly after we review Hertz's specific set of proposals.[5]

But let me again reiterate that the multiscalar strategies that operate so effectively within mechanics should not be regarded as a universal philosophical panacea. Different areas of study generally demand the close mastery of significantly different collations of foundational detail. With respect to, e.g., the delicate nuances of ethical discourse, I would presumptively concur with J. L. Austin's sentiments in this celebrated passage from "A Plea for Excuses":

> [O]ur common stock of words embodies all the distinctions men have found worth drawing, and the connections they have found worth drawing, in the lifetimes of many generations: these surely are likely to be more numerous, more sound, since they have stood up to the long test of the survival of the fittest and more subtle, at least in all ordinary and reasonably practical matters, than any that you or I are likely to think up in our arm-chairs of an afternoon—the most favored alternative method. (Austin 1966, 182)

But the moral wisdom embodied within the contextualized discriminations that Austin elucidates is most likely regimented within linguistic encodings that bear little structural resemblance to the multiscalar machinery that we shall study in detail, beyond a shared reliance upon contextualized discriminations. And I would further recommend that we disassociate our diagnostic ambitions from the "therapeutic" suggestions that one sometimes finds in Austin (and, more frequently, the later Wittgenstein) to the effect that "conceptual muddles" comprise deviant circumstances in which pre-established "rules of usage" or "grammar" have been improperly applied. I believe that the term "mistake" rarely captures an apt assessment of the dilemmas we confront in philosophy. Our evolved descriptive policies present us with undeniable inferential opportunities that have become actualized within our linguistic policies in imperfect manners (just as Darwin's biological adaptations rarely prove optimal in their fulfillments). As such, no "rules" that can be justifiably classified as "violated" appear as clearly marked within the established practice. In the same vein, learning the procedures whereby a novel inferential routine can be successfully implemented is not the same as adequately appreciating how the routine operates. We can execute quite complicated feats on a computer without any real understanding of the information-processing

---

[5] More generally, the stark opposition that Carnap draws between "metaphysics" and "anti-metaphysics" continues to dominate analytic philosophical thinking to this present day, ignoring the "small metaphysics" concerns we shall highlight in our examination of Hertz.

involved. So any Austin-like study of "excuses" must confront the further question of why these nuanced discriminations are in fact useful.

Nonetheless, it is surely a symptom of methodological attitudes gone awry when perceptive investigators such as Austin become commonly disparaged by the ersatz rigorists of today as "unscientifically unsystematic." Such "formalist" critiques should bear in mind W. K. Clifford's admonition that "it is so much easier to put an empty room tidy than a full one" (Clifford 1901, 2). When an evolving linguistic architecture becomes gradually cemented in place through a progression of "Euler's pencil" improvements, the accommodating "rooms" are apt to prove over-stuffed in the manner of a Victorian parlor. To presume otherwise relies upon an excessively antiseptic conception of science's developmental processes. But our inherited Carnapian traditions of "formal explication" embody exactly such a presumptive mistake; they draw a faulty equation between "being scientifically rigorous" and "practicing formal methods" that seriously inhibits useful philosophical research to this day, even with respect to topics far removed from "science" per se. An adequate unraveling of the "small metaphysics" puzzles of which I write typically demands suppler resources of diagnostic thought than such a blinkered view of "science's methods" encourages.[6]

This book's title stems from a conversation I once had with Robert Pippin with respect to Douglas Sirk's film *Imitation of Life*. The characters in this movie suffer from various circumscribed conceptions of "what it is to live a fulfilling life," drawn from the social frameworks from which they come. As a result, each pursues stunted "imitations" of the goals they should more reasonably seek (Sirk reported that he agreed to direct the film largely on the basis of its Fanny Hurst-derived title). Pippin captures this theme of conventionalized entrapment as follows:

[Their] situation borders on the tragic, not the melodramatic. [None] of them can fix [their regimented social expectations] and there is no acceptable way to resign [themselves] to it and live with it. The blindness, the strategic blindness, we have often seen in Sirk, is the only subjective way to live with such a hopeless, objective situation.... [T]hese are all cases of wrong without, however paradoxical it is, there being any wrongdoers.   (Pippin 2021)

Without intending to be snarky (I was mainly seeking an excuse for hijacking the film's poster as cover art!), I do believe that excessive devotion to the ersatz standards of "rigor" criticized in this book has likewise compromised contemporary philosophy's abilities to deal with the deeper forms of conceptual mystery at a diagnostic level commensurate with the problems themselves.

---

[6] Baz (2012) provides a valuable overview of contemporary dismissals of Austin-like views in this manner, although I feel that Baz is insufficiently sensitive to the concerns with recursive learnability that lie in the background of such opinions. My own appeals to developmental "seasonalities" in Chapter 8 represent an attempt to steer a middle course between these diagnostic extremes.

2

Prospectus

The lady upon hearing this embraced the [tutor] and said: "It is easy
to see, sir, that you are the wisest man in the world. My son will be
entirely indebted to you for his education. I think, however, it would
not be amiss if he were to know something of history."

"Alas! madam, what is that good for?" answered he; "there certainly is no useful or entertaining history but the history of the day;
all ancient histories, as one of our wits has observed, are only fables
that men have agreed to admit as true. With regard to modern
history it is a mere chaos, a confusion of which it is impossible to
make anything. Of what consequence is it to the young marquis, your
son, to know that Charlemagne instituted the twelve peers of France
and that his successor stammered?...A man of quality, like the
young marquis," continued he, "should not rack his brains with
useless sciences."...

The young man's brain was soon turned; he acquired the art of
speaking without knowing his own meaning, and he became perfect
in the habit of being good for nothing.

Voltaire (Voltaire 1901, 7)

(i)

In my estimation, the philosopher/physicists of the later nineteenth century
comprise a great epoch within philosophical thinking, a period whose innovative
suggestions were prompted by an intriguing combination of close empirical
observation and tactical ingenuity. They realized that raw armchair "intuitions"
and metaphysical doctrine, as the latter had become consolidated in their
time, were serving as roadblocks to improved science. This recognition led to
insights with respect to sound methodology that are worth reviving today.
However, to profit from their wisdom, a significant body of faulty folklore with
respect to "classical mechanics" needs to be corrected, as it obscures the subtleties
of the conceptual difficulties they confronted. Writers commonly patronize
this era as a time of settled complacency, marred only by a few residual "clouds"

Imitation of Rigor: An Alternative History of Analytic Philosophy. Mark Wilson, Oxford University Press. © Mark Wilson 2022.
DOI: 10.1093/oso/9780192896469.003.0002

of doctrinal uncertainty.[1] But this characterization is misleading, for these philosopher/physicists displayed an innovative willingness to rethink old shibboleths in manners from which we can still profit, especially in light of the fact that academic philosophy has imprudently returned to many of the repressive metaphysical assumptions against which they rebelled.

It may strike most of my readers as scarcely credible that current research in, e.g., ethics or metaphysics, might benefit from a reconsideration of various old-fashioned concerns with respect to Newton's Third Law of Motion. Nonetheless, such interconnections persist, for the byways of philosophy meander through broad landscapes in which current percepts remain significantly shaped by the half-forgotten debates of prior times. Contemporary philosophical argument is conducted by handed-down anecdote to a surprisingly large degree, especially in circumstances that involve scientific methodology in some manner or form. These folkloristic narratives supply academic scholars with a cast of stereotyped heroes and villains, embedded within a retrospective mythology that tidily illustrates how right supposedly triumphed over wrong. Much of this dubious lore involves "classical physics" largely because the subject appears to adopt a simple point of view that we can all absorb readily. As a result, modern writers invoke incidents from our Newtonian past as liberally as eighteenth-century writers drew upon Herodotus.

Except in appendix and footnote, this monograph will not attempt to remedy these historical misrepresentations. But I am obliged to outline the basic intellectual dilemmas that Hertz confronted to the minimal degree required to appreciate the diagnostic skills that he displays in his *Principles*. To me, these admirable insights should stand as paragons of the informal investigative rigor that we should strive to emulate. The present chapter and its successor will be devoted to fleshing out Hertz's scientific context enough to appreciate the conceptual dilemmas he faced and the philosophy of corrective repair that he applied to these problems. As a side benefit, we will review various insufficiently appreciated details that further exemplify the "small metaphysics" concerns that motivated Leibniz, Kant, du Châtelet, and the other great philosophers of earlier eras. Chapter 4 will discuss Hertz's particular contributions to these "small metaphysics" issues.

However, skilled diagnosis and suitable remedy do not inevitably coincide. In a reconstructive proposal that seemed entirely reasonable at the time, Hertz attempted to correct the conceptual tangles he uncovered (which were impeding physics's advance in a manner we shall discuss) by formulating a careful and self-consistent axiomatics in the manner of Euclid. In doing so, he also hoped to wrap

---

[1] This smug appraisal is often attributed to Lord Kelvin on the basis of Thomson (1901). The actual contents of this essay reveal his willingness to rethink fundamental physical doctrine in quite radical ways.

the subject within a protective cocoon that could withstand the invidious invasions of "intuitive" misgivings or ill-considered metaphysics. As we later see, these prophylactic purposes have played a significant role in shaping later philosophical thinking with respect to "rigor" and the proper scope of "metaphysics."

However, recent developments within applied mathematics have revealed that Hertz's axiomatic cure does not represent the corrective wanted and that a so-called "multiscalar architecture" provides a significantly superior remedy. We shall outline the alternative tactics involved in Chapter 5. In our concluding chapters, we will subsequently survey several promising arenas in which allied forms of liberalized thinking with respect to supportive "architecture" might relieve the tensions which bedevil current discussions of "causation," "realism," and "truth."

The chief moral I will derive from the historical developments just sketched concerns the manners in which later generations of "ersatz rigorists" have extracted unwarranted methodological "lessons" from Hertz's attempts to supply axiomatic correctives. The very notion of "axiomatic presentation" in itself is not especially problematic. On the contrary, employed with proper discretion, strict axiomatics provides an extremely valuable tool for evaluating important aspects of conceptual complexity. In fact, David Hilbert became interested in the topic partially through an acquaintance with Hertz (1894), and his own work in geometry inspired valuable research within two basic categories. In the first, Hilbert's school eventually focused our understanding of "axiomatic rigor" in the direction of "first-order consequence of sentences α" in the manner generally adopted today. As a secondary benefit, Hilbert placed the task of improving upon Hertz as the sixth of his famous set of problems that mathematicians should address in the years to come (Hilbert 1902). Through his student Georg Hamel and others who followed thereafter, Hilbert's suggestions served as a prod that eventually produced the modern consolidation of continuum mechanics to be discussed in the appendix to Chapter 3, reflecting the great improvements in our modern understanding of the subject.[2] Much of this clarificatory work was accomplished in an axiomatic vein, a topic to which I'll return in the same appendix. One of the historical complications that we will need to circumvent in our own discussion stems from the fact that no one in Hertz's time knew how to articulate a direct approach to classical continuum mechanics in a mathematically coherent way, despite the fact that the subject supplies a better housing for the basic ideas of the subject. In the final analysis, these difficulties do not affect the

---

[2] Unfortunately, the lack of interest that physicists took in classical mechanics after the rise of quantum mechanics (which I will outline later) led to a situation where Hilbert's 6th problem became poorly understood, even in seemingly "expert" presentations such as Wightman (1983) and Corry (2010). The fact that Hilbert's own interests shifted thereafter to the novelties of General Relativity didn't help either.

Third Law conceptual tensions with which Hertz was centrally concerned in any vital way, but we will need to discuss some of the continuum mechanics obstacles he faced in order to appreciate some of the peculiar tactics that Hertz adopts in working out his proposed axiomatics. A greater expositional problem arises from the fact that many of his motivating problems have simply become forgotten about, having become displaced by a simplistic formalism that I shall articulate as "Euler's recipe" in Chapter 3. This significant "loss of problems" has led most analytic philosophers into mistakenly presuming that the doctrines of "classical mechanics" can be neatly accommodated within a simple axiomatic format. In the next two chapters we will need to correct some of these misapprehensions before we can properly appreciate the conceptual dilemmas that Hertz correctly identifies (as provided in Chapter 4).

Returning to the narrow issues of axiomatic organization per se, the two forms of formal improvement mentioned above constitute genuine advances in our understanding, and nothing in this book suggests otherwise. The genuine harms that I shall document entirely arise from philosophical misapplications that stem from two central sources: (1) the would-be rigorist's propensity for rushing headlong into axiomatic stipulation without adequate diagnostic reflection; and (2) the unhelpful disposition to substitute the logical features of "axiomatized theories" for the more nuanced classifications that applied mathematicians have developed for similar discriminatory purposes. Both of these propensities had become firmly locked in place by the time that Rudolf Carnap announced his "anti-metaphysical" program in the later 1920s and have continued uncorrected to the present day (even within the philosophical schools that favor reviving "metaphysics" in an ersatz rigorist vein).

In writing of these logic-inflected classifications, I particularly have in mind the widespread abuse of "initial condition," "boundary condition," and similar vocabulary within philosophy of science circles, a circumstance that often engenders a faulty illusion of precise classification. When I complain about the distortions of "theory T thinking" in the sequel, I frequently have these specific terminological misfires in mind. However, I don't wish to devote large expanses of text to rectifying these misclassifications here (for such topics, see Wilson (2017)). On rare occasions, I will sometimes insist upon stricter readings of, e.g., "initial conditions" in order to highlight certain features of our "multiscalar" alternatives in firmer terms. Nonetheless, our central focus will not fall within the narrow corridors of descriptive philosophy of science per se but will instead concentrate upon the ways in which substantive conceptual confusions can creep into an otherwise beneficial form of linguistic activity and improve our appreciation of the tacit safeguards that prevent these intrusions from occasioning great harms.

Much of this "ersatz rigor" misdirection stems from the simple fact that a rudimentary understanding of first-order logic is easier to achieve than a comparable grasp of differential equations, and many philosophers have happily substituted logical distinctions for the more demanding classifications of applied mathematics. By incorrectly presuming that axiomatic resolution in Hertz's manner can successfully resolve the problems for which this remedy was originally intended, twentieth-century authors such as Rudolf Carnap posited that the run-of-the-mill "theories" encountered in everyday science could be advantageously "idealized" as fully axiomatized "theories T." Here is a characteristic appeal to such a "convenient fiction":

> To determine the direction in which physics at any given stage should move forward, the *fiction of a completed construction of physics* can be very useful, as a kind of goal in the infinitely long run.   (Carnap 1923, 221)

This policy of invoking logicized and thoroughly regimented replacements continues to be regarded as a "harmless idealization" that can ably streamline methodological discussion. For what harm can occur if the incipient "theory" in question will eventually submit to Hertzian rigorization later in its career? By accepting these fictional completions, philosophers believe that they can freely substitute coarse classifications of a vaguely logicized ilk for the careful discriminations of the applied mathematicians, leading to inadequate conceptions of "laws" and "initial and boundary conditions" that can substantially impede accurate philosophical diagnosis. These distortions supply the background assumptions that allow David Armstrong (as cited in the previous chapter) to unapologetically presume that he can usefully pronounce upon "laws of nature, disentangled from confusing empirical detail" (Armstrong 1983, 6). Through these simple incantations, the developmental quirks characteristic of a scientific language's evolving linguistic adjustments become concealed from view, and all of the delicate grain characteristic of real-life conceptual conflict are sanded away into bland "theory T idealization." I hope that our retrospective review of the concrete riddles with which Hertz was originally concerned will vividly illustrate the follies of excessive schematism of this character.

Familiar psychological factors encourage these same modes of thematic rigidification. Every academic investigator would like to become celebrated as the originator of some grand thesis or other, whether in science or outside of it. If proper "theory" demands that one's observations should be jammed into the ill-fitting suit of quasi-axiomatized doctrine, so be it. But these policies of artificial regimentation are alien to the normal character of intellectual advance, for the "potentially useful suggestions" of real life "theory" that require gentle augmentation by many hands. I believe that philosophical diagnoses of conceptual disharmony can proceed along more profitable trajectories if their practitioners recognize that the generating sources of their puzzles might well lie within descriptive ingredients that fail to fit together as an integral whole, without these incongruities making themselves evident in any component doctrine. As a result, it can easily happen that no hint of conceptual irregularity affects the core ingredients that philosophers typically regard as "theories" and can only be found within the tiny reforgings actually responsible for the subject's slowly accumulating practicalities. In such cases, the primary focus of our conceptual detective work must attend to these telltale patterns within the applicational matrix, rather than staring blankly at the unrevealing façades of prettied up "theories." So the "harmless policy" of presuming that this underlying roughness has been successfully smoothed away into an acceptable "theory T" format does not constitute a "useful epistemological fiction" but inadvertently locks into place an ill-advised unwillingness to scrutinize the irregular courses of language adjustment at the requisite level of revealing detail. I believe that allied excuses for ignoring complexity in the name of "ersatz rigor" have compromised many fields of current philosophical endeavor, even in circumstances where thoughts of "axiomatization" per se have never crossed their investigators' minds.

In the present book, we shall largely concentrate upon the historical developments that trace how these narrowed standards of philosophical contemplation flowered into methodological dominance under the formalist encouragements of Rudolf Carnap and his optimistic cohort of self-styled "logical empiricists." The axiom schemes "themselves" have largely disappeared from prominence, but the "rigorist" propensities behind them have lived on, afterwards, in the form of "theory T" presumptions with respect to the viable architectures of effective reasoning and explanation. Our ongoing pencils frequently prove wiser than ourselves, and we must learn to recognize their shaping ministrations wherever they appear.

(ii)

Hertz's approach to the conceptual difficulties he confronts reflect a critical decision point within his endeavors. On the one hand, his thinking about the evolutionary sources of conceptual disharmony within an improving practice are similar to my own, as outlined in our opening chapter. As such, allied views were widely shared by other philosopher/scientists of his era, as they struggled to rationalize the startling innovations that were enhancing science everywhere in the face of conservative "metaphysical" objections to their implementation (Wilson 2020). In what follows, I shall often cite Ernst Mach with respect to the conceptual liberties desired within science, as he expands generously upon these topics while Hertz remains more tersely circumspect. Both writers plainly oppose "conceptual analysis" in the mode of internalist investigation that remains popular in analytic philosophy circles today and instead contend that a living descriptive practice should be investigated as an integrated piece of holistic machinery, in order to pinpoint the misfunctioning locales where poorly matched gear wheels grind against one another. I believe that this view of this conceptual remedy appealed strongly to Ludwig Wittgenstein, but this functional approach to philosophical diagnosis has been largely expunged from our contemporary scene by the misguided demands of "ersatz rigor." Accordingly, the present book should be regarded as an attempt to resuscitate Hertz's original theses with respect to the developmental origins of conceptual disharmony.

But the second aspect of Hertz's work that will concern us in this book traces to the presumption (entirely reasonable in terms of what was known at the time) that the proper instrument of ameliorative repair should assume the form of a self-consistent axiomatization. Hertz didn't live long enough to confront the fact that the scheme he eventually proposes along these lines is thoroughly untenable in the demands it places upon rigorous mechanical reasoning. By Carnap's time, the unworkable impracticality of Hertz's axiomatization had become widely acknowledged, but a severely simplified view of "classical mechanics" simultaneously arose to prominence which readily submits to a rather trivial axiomatization of a "theory T" character (viz., the "Euler's recipe" scheme of Chapter 3). As a result, Hertz's insights into the nature of conceptual repair were largely forgotten, whereas his overly optimistic faith in axiomatic remedy soldiered on (in the guise of mistaken presumptions that suitable replacements had been found). Unfortunately, many of these logical empiricist misconstruals of Hertz's intentions have remained with us to the present day.

For these reasons, I will attempt to correct these misapprehensions as efficiently as possible in our opening chapters. As a consequence, I must respond in some minimal way to these historical misconceptions within my opening chapters. The remainder of the present section corrects some common misunderstandings with respect to what should comprise "the basic objects of classical physics." Chapter 3 then reviews the particular manner in which a seemingly insignificant bit of gnarly behavior becomes an issue of considerable concern for Hertz and the physics community of his time. Only in Chapter 4 (under the heading of "The Mystery of Physics 101"), will we be truly ready to study Hertz's identification of the specific "gnarly behavior" in question, which I regard as a paragon specimen of how a perceptive conceptual analysis should be conducted. Some interpretative exegesis is required on this score because of the daunting aspects of Hertz's prose, which is often condensed to the point of incomprehensibility. So I shall first attempt to reformulate his basic motivations in more amical terms. Insofar as I fail in these objectives, the reader should simply skip past any refractory passages and regain the narrative thread in the pages ahead. In Chapter 5, we will finally consider a modern, non-axiomatic resolution of the difficulties that Hertz identifies (viz., the multiscalar modelings previously mentioned). This approach applies a more sophisticated conception of linguistic architecture to Hertz's difficulties, a tactic that better accords with everyday practice and the general picture of linguistic evolution outlined in Chapter 1.

These suggestions also provide concrete exemplars of the broadly structuralist approach to philosophical exegesis that is possibly found in Wittgenstein's writings and was definitely endorsed by J. L. Austin and other figures of his generation. I believe that the valuable insights they offered have become significantly underappreciated in recent years due to their lack of "ersatz rigor" gloss. With respect to genuinely rigorous articulation, the multiscalar architectures of modern computer science have been subjected to the tightest standards of mathematical scrutiny and can easily "out-science" the semi-scientific proposals of the ersatz rigorists any day of the week. This is not to say that good philosophy must replicate these highly disciplined standards of machine implementation, but it can still delineate unexpected reasoning architectures that are implemented within our everyday modes of discourse. I suspect that Austin himself might have been pleased by this redemptive analogy, for he clearly conceived of his own proposals as exercises in "linguistic engineering" of a sort.

As Voltaire's tale of the sardonic tutor indicates, sequences of complex historical events are afterwards condensed by subsequent generations into simpler narratives, upon which easy morals can be conveniently hung. This is certainly the fate that has befallen our modern appreciation of both *The Principles of Mechanics* and "classical physics" generally, at least within popular culture and among a fair number of contemporary physicists.[3] This after-the-fact thinning of problems has created a situation in which Hertz's original responses to a serious form of conceptual quandary have become significantly misconstrued due to doctrinal amnesia with respect the motivating difficulties. But glib resolutions of simplified dilemmas often encourage wrongheaded solutions. Philosophical presumptions to the effect that "everybody understands what 'classical mechanics' comprises" significantly promote misleading conceptions of scientific methodology.

All the same, I do not want this volume to degenerate into a lengthy disquisition upon the complexities of classical doctrine, for the general philosophical audience for whom this volume is intended would gain very little from that. All that I hope to extract from Hertz himself is a finer appreciation for the manner in which an evolving practice can organically generate tiny conceptual puzzles in the courses of its practical improvements. What first appears as a minor irregularity can turn out to be symptomatic of an unnoticed "architectural" feature of the wider doctrinal framework to which the little "irregularity" belongs (in the manner of the stray thread that unravels the entire sweater when pulled). This realization should make us wonder, "How might a descriptive practice remain viable while it remains infected by a problematic virus of the sort Hertz correctly identifies?" Pursued in

---

[3] Mechanical engineers are less susceptible to such stereotypes, largely because they must rely upon tradition's richer bag of descriptive tools in their daily activities.

this vein, our multiscalar alternatives provide interesting suggestions with respect to the exact roles that mathematics should play within our descriptive enterprises. Some of these considerations will be discussed in our concluding chapters. As noted earlier, many of the grander examples of historical metaphysical speculation drew their motivational roots from "small metaphysics" concerns of this general "how does mathematics aid physical description?" character. At present, however, the soothing assurances of "ersatz rigor" have directed our attention away from these delicate issues of data registration and have focused instead upon the cruder conceptions of metaphysical endeavor that we inherit from Carnap and his generation.

In the remainder of this chapter and the next, we'll review enough of the background required to appreciate the diagnostic dilemmas Hertz faced, without fully exploring the complicated nooks and crannies that surround his developed doctrines (some of the latter will be surveyed in the appendix to Chapter 3 labeled "Historical Complexities").

(iii)

Let us now distinguish five basic classes of physical systems that one encounters within a general classical mechanics context:[4] (i) swarms of point particles that carry finite masses and serve as the sources of action-at-a-distance forces in the manner of standard celestial mechanics; (ii) disassociated rigid bodies such as billiard balls that interact with each other through both direct contact and action-at-a-distance forces; (iii) continuously contacting rigid bodies in the fashion of a gear chain or clockwork; (iv) flexible extended continua such as rubber balls or violin strings; (v) continuous fluids. Items (i), (ii), and (iii) comprise *finite degree of freedom* systems in that they only require a finite number of descriptive parameters to fix their state, whereas (iv) and (v) demand *infinitely many parameters* to fulfill the same task given their abilities to distort into any continuous

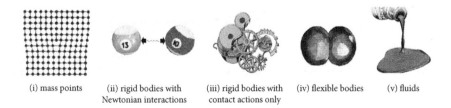

(i) mass points   (ii) rigid bodies with   (iii) rigid bodies with   (iv) flexible bodies   (v) fluids
                   Newtonian interactions   contact actions only

---

[4] Hybrid approaches involving a mixture of ingredients are possible as well (the upper-scale billiard ball treatment of Chapter 5 is a case in point), but we will ignore these complications at present.

shape we'd like ((iii) and (iv) are conventionally called "continua" or "fields" as a result) (Wilson 2013).[5]

For the sake of lexical brevity (but not historical accuracy), I will often call action-at-a-distance forces "Newtonian" in the sequel, despite the fact that Newton himself was equally partial to contact forces between rigid bodies as well (Wilson 2021).

Although earlier generations had supposed that the fundamental primary objects in the universe might consist in bare point masses, this point of view had become firmly rejected by the epoch in which our philosopher/scientists operated.[6] To appreciate this rejection, we might cite J. C. Maxwell's wonderful entry on "Atom" from the celebrated 9th edition of the *Encyclopedia Britannica*. He begins by insisting that the atoms and molecules of real life need to be treated as flexible continua (iv) in the manner of bells:

> We may compare the vibrating molecule to a bell. When, struck, the bell is set in motion. This motion is compounded of harmonic vibrations of many different periods, each of which acts on the air, producing notes of as many different pitches. As the bell communicates its motion to the air, these vibrations necessarily decay, some of them faster than others, so that the sound contains fewer and fewer notes, till at last it is reduced to the fundamental note of the bell.... A spectrum of bright lines, therefore,

vibrating snape

---

[5] Mathematically, the first group cobbles by with *ordinary* differential equations, whereas the continua groups require *partial* differential equations (or something fancier along the same lines). A chief error in theory T thinking lies in not distinguishing the conceptual demands of these two settings adequately. Many of the observations I will offer in this book rest upon those absent discriminations, although I shall largely avoid such technicalities as best I can.

[6] Point masses were actively endorsed as real physical entities within Roger Boscovich's (1758) work *Philosophiæ naturalis theoria*, and similar views were advocated by some of the French atomists of the early nineteenth century. By Hertz's time, this naive approach was generally dismissed as a "molecular cloud-land" (Thomson 1901), although a previous generation of Scottish physicists had regarded Boscovich's doctrines more favorably (Olson 1975). P. G. Tait comments:

> [Boscovich's] theory must be regarded as a mere mathematical fiction, very similar to that which (in the hands of Poisson and Gauss) contributed so much to the theory of statical Electricity; though, of course, it could in no way aid inquiry as to what electricity is. (Tait 1890, 21)

Nevertheless, point masses continue to play a prominent introductory role within these authors' own work (Thomson and Tait 1912) precisely because of the measure-theoretic problems surveyed in Chapter 3. As that chapter also explains, point masses made an unexpected comeback in the 1920s due to the transitional mathematical bridge of so-called "quantization." These factors have encouraged a modern propensity to read authors like Kelvin or Hertz in anachronistic point mass terms, which severely distorts their intended meanings (Hertz did not take point mass mechanics seriously enough to include it as one of his possible "images" of mechanics). Insofar as I can determine, a fair number of the readings of Hertz found in Baird, Hughes, and Nordmann (1998) commit this mistake or something closely akin (in fairness, some of Hertz's somewhat odd terminology—viz., "connections between material points"—can suggest as much, although I will provide their proper reading in the appendix to Chapter 3). The best commentary on *Principles* is Lützen (2005), upon which I shall frequently rely.

indicates that the vibrating bodies when set in motion are allowed to vibrate in accordance with the conditions of their internal structure for some time before they are again interfered with by external forces....

Every molecule, so far as we know, belongs to one of a definite number of species. The list of chemical elements may be taken as representing the known species which have been examined in the laboratories. Several of these have been discovered by means of the spectroscope, and more may yet remain to be discovered in the same way. The spectroscope has also been applied to analyze the light of the sun, the brighter stars, and some of the nebulae and comets, and ... a considerable number of coincidences have been traced between the systems of lines belonging to particular terrestrial substances and corresponding lines in the spectra of the heavenly bodies. We are thus led to believe that in widely-separated parts of the visible universe molecules exist of various kinds, the molecules of each kind having their various periods of vibration either identical, or so nearly identical that our spectroscopes cannot distinguish them. We might argue from this that these molecules are alike in all other respects, as, for instance, in mass. But it is sufficient for our present purpose to observe that the same kind of molecule, say that of hydrogen, has the same set of periods of vibration, whether we procure the hydrogen from water, from coal, or from meteoric iron, and that light, having the same set of periods of vibration, comes to us from the sun, from Sirius, and from Arcturus....

Atoms have been compared by Sir J. Herschel to manufactured articles, on account of their uniformity.... It seems ... probable that he meant to assert that a number of exactly similar things cannot be each of them eternal and self-existent, and must therefore have been made, and that he used the phrase "manufactured article" to suggest the idea of their being made in great numbers.

(Maxwell 1965, 445–84)

Here's Maxwell's argument in brief: (1) to possess discrete spectra, molecules must represent flexible continua akin to bells; (2) but bells can display an infinite variety of spectra depending upon their exact shapes; (3) in our universe, however, we only encounter a limited variety (less than one hundred known in Maxwell's day) of these molecular "bells"; (4) the most plausible explanation is that these specific forms of "bell" are of Divine manufacture. As such, this is the most compelling form of cosmological argument that I've encountered, whose counter-response requires the Pauli exclusion principle from quantum mechanics!

In any case, Maxwell's atoms and molecules are flexible continua capable of retaining their internal matter while rapidly altering their shapes in the manner of a bell.[7] Further speculations on Kelvin's part (which Maxwell regarded as

---

[7] Maxwell also favorably discusses Kelvin's more reductive "vortex atoms," which represent whirlpool singularities within a surrounding etherial fluid.

promising) suggests that these flexible (iv) atoms might in turn represent vortex singularities within an ambient fluid such as (v). However, as soon as Maxwell or Kelvin attempt to delineate the physical principles that govern these specimens of continua in an explicit manner, we find them oddly appealing to hypothetical swarms of isolated point masses and/or completely inflexible rigid bodies, despite the fact that neither scientist believed that they enjoyed a real existence. But why did they do this? Why didn't they directly specify the physics relevant to the flexible bodies that they thought existed? Today we recognize that this disparity largely stems from technical barriers, for the mathematical resources required to articulate continuum physics behaviors in a directly coherent way had not yet been formulated (as already indicated, the motivations for developing these tools came from David Hilbert's instigation). In the appendix to Chapter 3, I'll briefly outline the mathematical obstacles involved and how they were later overcome in the twentieth century.

As often happens, philosophical mottos step in where mathematics fails to tread. In the case at hand, a dubious methodology of "essential idealization" appears as a bridging principle, in the vein articulated by the Italian geometer Federico Enriques:

> [I]it seems well to begin the study of mechanics by taking up "material particles" as an elementary case. This fictitious construction is fully justified by its conformity with those principles for the simplification of scientific research which P. Volkmann calls principles of the isolating and superposing of phenomenal occurrences.   (Enriques 1914, 252)

Or, more fully, by the British statistician Karl Pearson:[8]

> If we take a piece of any substance, say a bit of chalk, and divide into small fragments, these still possess the properties of chalk. Divide any fragment again and again, and so long as a divided fragment is perceptible by aid of the microscope it still appears chalk. Now the physicist is in the habit of defining the smallest portion of a substance which, he conceives, could possess the physical properties of the original substance as a *particle*. The particle is thus a purely conceptual notion, for we cannot say when we should reach the exact limit of subdivision at which the physical properties of the substance would cease to be. But the particle is of great value in our conceptual model of the universe, for we represent its motion by the motion of a purely geometrical point. In other words, we suppose it to have solely a motion of translation . . . ;

---

[8] Although Pearson later gained his fame as a statistician, his early writings were devoted to continuum mechanics. For developed complaints with respect to invocations of "idealization" in this manner, see Wilson (2017, 139–40).

we neglect its motions of rotation and strain.... What right has the physicist to invent this ideal particle? He has never perceived the limited quantity, the *minimum esse* of a substance, and therefore cannot assert that it would not produce in him sense-impressions that could only be described by the concepts spin and strain. The logical right of the physicist is, however, exactly that on which all scientific conceptions are based. We have to ask whether postulating an ideal of this sort enables us to construct out of the motion of groups of particles those more complex motions by aid of which we describe the physical universe. Is the particle a symbol by aid of which we can describe our past and predict our future sequences of sense-impressions with a great and uniform degree of accuracy? If it be, then its use is justified as a scientific method of simplifying our ideas and economizing thought.

(Pearson 1892, 335–6)

Although Hertz himself is primarily interested in the behaviors of inherently continuous materials (such as the ether), he likewise presumes that the foundations of classical mechanics must be first codified in the behavior of finite degrees of freedom systems of the types (i)–(iii) (as we'll later see, Hertz hopes to reach the continuous behaviors by taking limits over rigid systems of type (iii) rather than point masses in Pearson's manner). This technical exclusion of continua as "foundational" generates an awkward situation in which the most natural means of resolving the conceptual tensions that troubled Hertz appeal directly to the notions of *internal pressure* and *stress* that only continuous flexible bodies can sensibly exemplify (when we examine Newton's Third Law of Motion in Chapter 4, we shall see why this is so). As it happens, Hertz himself wrote some important papers in which the internal stresses within a compressed ball are calculated, without being able to tell us what the term "stress" actually signifies, due to his lack of the mathematical concepts required (Hertz 1881). As a result, in the Victorian commentaries we shall examine, unformulated continuum physics doctrine looms as the unacknowledged elephant in the room, displaying the behaviors to which our philosopher/scientists would like to appeal but can't see how they can legitimately do so.

For our own purposes, we needn't descend into this somewhat daunting morass, beyond the bare recognition that untamed continuum behaviors lurk in the conceptual woodwork. And there is a second species of physical doctrine of which we must also be cognizant, despite the fact that Hertz never considers it as a live possibility. Unlike the continua, this omission isn't for lack of mathematical resources, but merely because Hertz views the format as too jejune to merit serious consideration. I here have in mind the stark framework that only contains the entities listed under (i): unextended point masses that interact solely by action-at-a-distance forces (never by contact since points lack the requisite geometries for doing so). Articulating a coherent mathematical basis for monitoring entities of this exceptionally simple character is a much easier task (up to a point), and the tools required for doing so will be reviewed in Chapter 3 under the heading of "Euler's recipe." Although Hertz himself often writes of "systems of material points" in his own developments, he never has these Newtonian point mass ensembles in mind but instead refers to connected ensembles of the rigid type (iii) inventoried above (in the appendix to Chapter 3, I'll explain his somewhat misleading choice of terminologies in these regards). When he considers (and subsequently rejects) his first "image" of mechanics, Hertz is instead thinking of dualistic descriptive formats such as (ii), in which extended rigid bodies possessing rigid geometries act upon one another through both direct contact and action-at-a-distance forces (which I sometimes label as "Newtonian" forces). And it is exactly within this generous tolerance of distinct classes of interactive forces that Hertz correctly pinpoints the problematic disharmonies that will concern us in the sequel. As for the bare point mass scheme, Hertz (and virtually every other physicist of his generation) would have laughed and responded, "Oh, that's merely a naïve fancy that Boscovich and some of his followers once entertained, but we have come a long way from such hypotheses because of the behavioral facts that Maxwell cites in his essay on 'Atom'."

In saying this, we should not overlook the "never ending fascination" of the mathematics proper to point mass mechanics, which Lord Kelvin accurately characterized in this amusing manner:

[Before considering a problem within continuous dynamics], I want to touch upon the somewhat cloud-land molecular beginnings of the subject and refer you back to the old papers of Navier and Poisson, in which the laws of equilibrium or motion of an elastic solid mere were worked out from the consideration of points mutually influencing one another with forces [which are] functions of the distance. There can be no doubt of the mathematical validity of investigations of that kind and of their interest in connection with molecular views of matter; but we have long passed away from the stage in which Father Boscovich is accepted as being the originator of a correct representation of the ultimate nature of matter and force. Still, there is a never ending interest in the definite mathematical problem of the equilibrium of motion of a set of points endowed with inertia and acting upon one another with any given force. We cannot but be conscious of the one grand application of that problem to what used to be

Kelvin

called physical astronomy, but which is more properly called dynamical astronomy, or the motions of the heavenly bodies.   (Thomson 1904, 123)

The "definite mathematical problem" to which Kelvin alludes is the general behavior of ordinary differential equations of an evolutionary character (so-called ODEs), for which an enormous literature has been devoted (sometimes under the misleading heading of "the mathematics of classical mechanics"). In Chapter 3, we shall discuss some of the characteristic features of this greatly reduced scheme. I again observe that in our Victorian epoch, this approach had become widely rejected as inadequate to the demands of either elasticity or molecular physics.

For largely spurious reasons, this simple ODE framework made an unexpected comeback within the physics community (but not the engineering!) during the 1930s, largely because of the formal demands of quantum mechanics. In particular, the simplest forms of quantum mechanical model can be obtained by quantizing simple point mass models (viz., replacing ODE operators by Schrödinger equation surrogates). Under this transference, molecular swarms that are intrinsically unstable from a classical mechanics perspective become astonishingly stable and better suited to modeling actual molecular arrays. This parochial advantage has generated a pedagogical environment in which point mass mechanics becomes the central "classical physics" that a beginner in physics needs to acquire,[9] creating a false impression that its simple ODE provisos capture the amazing breadth of the subject as Hertz and his colleagues knew it.

In my estimation, this faulty impression has contributed significantly to the schematic philosophical conceptions of "how science works" of which this book complains. As a practical descriptive practice, point mass mechanics is woefully underequipped (which is why Hertz ignores it entirely). But its operations are easy to grasp, leading many philosophers to presume that they "fully understand how classical mechanics works" on the basis of a narrow and inadequate partial thread. These unhelpful misapprehensions multiply when the philosophical "understander" further presumes that other familiar forms of "classical behavior" (such as billiard ball collisions) can be neatly accommodated within the point mass schematism. Such careless elisions frequently make it impossible to determine what such an author is talking about (Wilson 2009).

---

[9] Upon this same point mass basis, the formalisms known as "Hamiltonian" and "Lagrangian" are commonly introduced as "theorems." As we shall see, Hertz does consider these descriptive tactics under the heading of his "second image."

(iv)

In any case, in his celebrated preface Hertz outlines three basic approaches that one might adopt in attempting to eradicate the foundational tensions within mechanical doctrine (he calls these three approaches "images" for reasons we needn't worry about now). The first of these "images" comprises the assemblages characterized as (ii) above: extended objects that act upon one another through action-at-a-distance forces and the local forces that arise between directly contacting surfaces. The latter are commonly called "constraint" or "traction" forces, although Hertz often resists labeling the latter as true "forces" at all (I will not follow this misleading practice). As Hertz fully realizes, a physicist's ability to appeal to both of these force types substantially contributes to her ability to capture an extremely wide range of physical systems with remarkable accuracy, to such an extent that most macroscopic systems can be successfully treated with these "classical" modeling tools alone (*modulo* the need to extend their reach to embrace continua as well).

Nonetheless, Hertz's chief complaint with respect to this generous first "image" is that these two classes of force enter its framework via distinctly different mathematical routes and wind up displaying subtly different characteristics as a result. Most of the time these slight disparities generate no problems, but in certain critical situations the disagreements explode into significant applicational dilemmas. For our purposes in this book, the specific conceptual disharmonies he identifies beautifully illustrate in microcosm the warring evolutionary factors that frequently engender allied puzzles within a wide variety of disciplines. And Hertz also identifies the basic reason why this happens. In gradually cobbling together an effective descriptive practice, we generally appropriate whatever useful reasoning tools we find close to hand, without verifying that they are fully in accord with the other assumptions we have already made. Worse yet, reflective cogitation may fail to reveal any conceptual miscreant that bears responsibility for these applicational clashes, for each party to the dispute may supply impeccable proofs of its practical centrality. We will later find that the doctrinal repairs required are more radical in their architectures than Hertz anticipated (his own axiomatic recommendations proved impractically stiff in these regards). As it happens, recent advances in computational strategy have opened up fresh avenues along which better forms of applicational reconciliation can be obtained, and these will supply us with the methodological morals developed in Chapter 5.

For our own narrow purposes, Hertz's discussions of his further foundational "images" needn't concern us much, but since their characteristics have been widely misidentified, I will briefly outline them here. His second "image" articulates a viewpoint that was once called "energetics," which divides the behaviors of an evolving system into kinetic and potential energy components (Helm 1898). The movements of a swinging pendulum illustrate these distinctions, as it continually shuttles energy between a purely potential storage (at the ends of its swings within a gravitational field) and its kinetic movements (as when the bob whizzes past its lowest

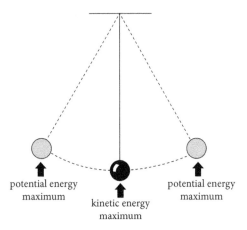

point). The manner in which this energetic transference occurs is governed by some form of variational principle applicable to the Lagrangian or Hamiltonian functions encountered within every modern textbook on mechanics.

Similar governing formulas can be derived from a force-based starting point such as (i) at the price of deriving the potential energy term (usually designated as "V") from some restricted set of underlying force laws, but Hertz does not adopt this reductive point of view in his discussion. He instead expects that a modeler can assign a suitable Lagrangian (or Hamiltonian) to a target system after ascertaining its possible configurations through direct experiment, by wiggling the system to uncover the directions in which it can be freely manipulated without falling apart. This manipulation data can allow a modeler to assign the system a suitable set of so-called "independent generalized coordinates" ϕ_1, ϕ_2, ϕ_3, ... in which any of these coordinate values can be varied without affecting the others. Once these independent determinations have been articulated, the modeler can assign an appropriate V to the generalized "location" <ϕ_1, ϕ_2, ϕ_3, ... > as its behaviors dictate, without any concern that a point mass force scheme exists to support such a V.[10] Indeed, Maxwell and Kelvin often utilized Lagrangian descriptions to sidestep unnecessary hypotheses with respect to molecular underpinnings.

Hertz's stated reluctance to embracing this wholly energetic route largely turns upon the format's difficulties in handling so-called non-holonomic constraints (such as rolling) in a natural way. We needn't explore these technical concerns in any depth here (Lützen 2005, 192–7).

His third "image" of mechanics is the one he himself favors, whose governing principles had not been previously formulated in the purist form he demands.

---

[10] Forces can be derived from such a V through a partial differentiation, but the results are not guaranteed to be of a Newtonian character. One sometimes encounters loose assertions to the effect that point mass, Lagrangian, and Hamiltonian formulations are "equivalent," but such claims are only sustainable against a restricted theoretical background.

However, the underlying intuitive conception is rather old: it is simply the notion of a clockwork universe, in which all interactions between component parts arise when one member slides smoothly across a neighboring surface. Arrangements of this kind are often called "linkages" and in 1875, the mathematician J. J. Sylvester articulated a portrait that is very close to Hertz's third image:

> It is quite conceivable that the whole universe may constitute one great linkage, that is, a system of points bound to maintain invariable distances, certain of them from certain others, and that the law of gravitation and similar physical rules for reading off natural phenomena may be the consequences of this condition of things. If the Cosmic linkage is of the kind I have called complete, then determinism is the law of Nature; but, if there be more than one degree of liberty in the system, there will be room reserved for the play of free-will.
>
> (Sylvester 1875, 214)[11]

To render this scheme operational, Hertz must find a governing law that can direct how kinetic energy will shuffle from one location to another within a complex mechanism. Most of the *Principles* is devoted to articulating the complicated apparatus he employs to this end (in the optional appendix to Chapter 3, I outline the unexpected tactics he employs in doing so). Without worrying about further details, a quick example can illustrate the challenges he confronts. Consider the "dwell point" mechanism illustrated. Let us inject a large portion of kinetic energy into the apparatus by turning the crank at C rapidly. As the crank turns, most of the time the weight at W will scarcely wiggle, but at a certain point in the cycle it will jiggle about rapidly, at which point the crank must slow in its own rotation so that

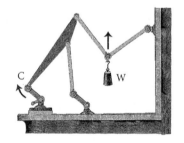

the total energy contained within the device is conserved. The axiomatization of mechanics he eventually proposes centers around the single "dynamic law" he proposes to this purpose.

---

[11]   I should remark that the watch-work illustrated is not a true exemplar of a proper linkage due to the inclusion of its mainspring, which introduces a disallowed notion of "spring force" into the picture. The system dynamics that Hertz requires should sustain the internal movements of a proper linkage forever through internal inertia alone, in the absence of any applicable friction. Descartes' vision of the physical universe was closely allied to Hertz's (Wilson 2021).

In any case, our own interests will focus upon his diagnosis of the evolutionary origins of his "first image" tensions with respect to "force" and his general recipe for their axiomatic amelioration, which can be characterized as "correction through doctrinal subtraction," viz., carve away unnecessary ingredients from an inconsistent totality until an internally coherent hull is reached. In my view, this represents the natural remedy one should first explore, but in the case at hand, correctives of this type wind up resembling the surgery that removes a wart but kills the patient as a side effect.[12] Indeed, Hertz's three "images" all suggest formalisms that involve unwelcome amounts of lower-scale speculation.[13] Are there alternative methods of cauterizing a localized descriptive defect without ruining the utilities of the larger practice to which it belongs? As just indicated, the multiscalar architectures surveyed in Chapter 5 do exactly that, and I hope that a warmer appreciation of their merits might liberate analytic philosophy from the straitened confines of "ersatz rigor" that have forced philosophical analysis into cramped quarters ever since axiomatic enthusiasms reigned back in Rudolf Carnap's heyday.

---

[12] A properly developed continuum mechanics does a much better job in avoiding such a large degree of immediate appeal to this otherwise unwarranted subterranean machinery, but as we've seen Hertz did not pursue this potential "fourth image." Consult the appendix to Chapter 3 for further details.

[13] Many of Mach and Duhem's animadversions against "metaphysical intrusions" into science stemmed from their thermodynamic concerns and were directed against what they regarded as an improper preference for the kinetic theory of heat as more "philosophically satisfying" (Wilson 2017, 136–47).

3

Inductive Warrant

What route would [our investigations] take? Several paths lay in sight;
the entrance to each was wide open and quite smooth; but hardly had
one gone along a path than one saw the causeway shrink, the track of
the route become unclear; soon one would see no more than a narrow
path half hidden by thorns, cut across by bogs, bounded by abysses . . . .
Where is he who would be carried through to the end desired, who,
one day, would come upon the royal way? . . . He who sows therefore
cannot judge the value of the grain; but he must have faith in the
fertility of the seed, in order that, without [faltering], he may follow the
furrow he has chosen, throwing ideas to the four winds of heaven.

Pierre Duhem (Duhem 1903, xl)

(i)

Within every human endeavor, its prevailing standards of "rigor" are commonly
tied to tacit preconceptions of which we are scarcely aware. Oliver Heaviside was
an idiosyncratic nineteenth-century engineer who developed outlandish innovations with respect to the manipulation of differential equations through brute
trial-and-error experimentation. With respect to the practical efficiencies of the
"operational calculus" he developed, Harold Jeffreys comments:

As a matter of practical convenience there can be no doubt that the operational
method is by far the best for dealing with the class of problems concerned. It is
often said that it will solve no problem that cannot be solved otherwise. Whether
this is true would be difficult to say; but it is certain that in a very large class of
cases the operational method will give the answer in a page when ordinary
methods take five pages, and also that it gives the correct answer when the
ordinary methods, through human fallibility, are liable to give a wrong one.

(Mahon 2017, 160)

But the inferential manipulations employed within Heaviside's calculus are very
strange and were sternly rejected by the mathematical rigorists of the period. In
the fullness of time (viz., 1950), adequate "semantic" underpinnings were finally

Imitation of Rigor: An Alternative History of Analytic Philosophy. Mark Wilson, Oxford University Press. © Mark Wilson 2022.
DOI: 10.1093/oso/9780192896469.003.0003

articulated by mathematicians, after a very careful scrutiny of the sources of its successes. In the interim, Heaviside justified his proposals as follows:

> A man would never get anything done if he had to worry over all the niceties of logical mathematics under severe restrictions; say, for instance, that you are bound to go through a gate, but must on no account jump over it or get through the hedge, although that action would bring you at once to your goal.... For it is in mathematics just as in the real world; you must observe and experiment to find the go of it.... All experimentation is deductive work in a sense, only it is done by trial and error, followed by new deductions and changes of direction to fit circumstances. Only afterwards, when the go of it is known, is any formal explication possible. Nothing could be more fatal to progress than to make fixed rules and conventions at the beginning and then go on by mere deduction. You would be fettered by your own conventions and be in the same fix as the House of Commons with respect to the dispatch of business, stopped by its own rules.   (Heaviside 1912, II 3, 279)

In compressed microcosm, the developmental vicissitudes of the operational calculus nicely illustrate the stealthy legwork required when evolving practice deposits infant mysteries upon our doorsteps.

With respect to *The Principles of Mechanics*, Hertz's prose is often evocative, but it is also notoriously terse, and the specific physical schemes he advances in the book are often strange and pursue unexpected gambits (although to a lesser degree than those of Heaviside). The book's introductory preface was widely regarded as inspirational among a large set of philosopher/scientists who grew up in the generation following Hertz's death, but the difficulties of parsing his words have also led to an enormous range of misunderstandings with respect to his actual methodological objectives. Indeed, his presentation commonly leaves his readers mystified as to why he chose to write the book at all, in view of the fact that he concentrates almost exclusively upon straightening out what the calls "the painful contradictions" one encounters in teaching basic mechanics:

> [I]t is exceedingly difficult to expound to thoughtful hearers the very introduction to mechanics without being occasionally embarrassed, without feeling tempted now and again to apologize, without wishing to get as quickly as possible over the rudiments, and on to examples which speak for themselves. I fancy that Newton himself must have felt this embarrassment when he gave the rather forced definition of mass as being the product of volume and density. I fancy that Thomson and Tait must also have felt it when they remarked that this is really more a definition of density than of mass, and nevertheless contented themselves with it as the only definition of mass. Lagrange, too, must have felt this embarrassment and the wish to get on at all costs; for he briefly introduces his

Mechanics with the explanation that a force is a cause which imparts "or tends to impart" motion to a body; and he must certainly have felt the logical difficulty of such a definition. I find further evidence in the demonstrations of the elementary propositions of statics, such as the law of the parallelogram of forces, of virtual velocities, etc. Of such propositions we have numerous proofs given by eminent mathematicians. These claim to be rigid proofs; but, according to the opinion of other distinguished mathematicians, they in no way satisfy this claim. In a logically complete science, such as pure mathematics, such a difference of opinion is utterly inconceivable.   (Hertz 1894, 6–7)

As we shall find in Chapter 4 ("The Mystery of Physics 101"), similar pedagogical discomforts have been expressed by a wide range of contemporaneous observers. But all of them (including Hertz himself) are certain that the wisdom they impart is "basically right," however muddled their introductory pronouncements may seem. Such attitudes create the impression that providing a satisfactory axiomatics for mechanics represents a rather prissy concern for "dotting one's i's and crossing one's t's" properly in these early developmental stages. Why should a great thinker like Hertz have devoted the final years of his short life to such an endeavor?

This impression of fusty irrelevance is deeply mistaken because pressing problems of what I shall call "inductive warrant" emerged within late nineteenth-century physics that appear unresolvable as long as Hertz's "rudiments of mechanics" remain hazy. We'll review these concerns in section (ii), but over the course of our discussion we will eventually find that the fuller dimensions of his motivating problems illustrate the "stray thread that unravels the sweater" phenomenon indicated earlier, in a manner that Hertz himself could not have anticipated. For the little "contradictions" that Hertz encountered within his pedagogy represent the tip of a much larger iceberg that encompasses "multiscalar architectures" and other unexpected gizmos of that character. But this is a basic feature of developmental life—our improving pencils frequently deposit us in unanticipated tangles of a foundational character. Their surprising innovations continually present us with the challenge of radically reconceiving the strategic underpinnings of our familiar reasoning patterns without spoiling the well-verified utilities of the practices thereby. And sometimes we silently accept these adjustments in expectation without recognizing that we have done so. Whether we do or not, we commonly find ourselves in intellectual circumstances that structurally resemble the quandaries of procedure that Oliver Heaviside confronted with respect to his peculiar "calculus." We may feel certain that some rational form of "semantic support" must underlie the bulk of our established reasoning policies, but we are darned if we know what it is. A biological analogy can be aptly cited on this score. We can justifiably presume that the peculiar foraging behaviors of various insects represent sensible adaptations to environmental demand, but we may utterly fail in identifying the strategic nature of these shaping imperatives.

In confusing circumstances of this character, we may temporarily lose our grip upon what the compasses of "rigorous reasoning" demand. At such times, it helps to wax "philosophical" in the old-fashioned sense of puzzling things over in a free-wheeling manner, with a sharp eye out for the revealing detail. As Charles Peirce once observed with respect to differential equations, "they do not divulge their secrets readily and one cannot charge at them like a knight in armor" (Peirce 1976, 25). But our current idols of "ersatz rigor" do not adequately encourage wide-ranging speculations of this sort. As such, the approach fails to anticipate the strategic rethinking we must continually exert as we struggle to improve our descriptive grip upon nature.

(ii)

In Hertz's case, his prompting motivations appear to have centered upon substantial concerns with respect to mechanical features of the "ether" through which Maxwellian electrical waves were supposed to propagate. Hertz supplies an accurate account of these issues in a lecture that he gave a few years prior to the *Principles*:

> What, then, is light? Since the time of Young and Fresnel we know that it is a wave-motion. We know the velocity of the waves, we know their length, we know that they are transversal waves; in short, we know completely the geometrical relations of the motion. To the physicist it is inconceivable that this view should be refuted; we can no longer entertain any doubt about the matter. It is morally certain that the wave theory of light is true, and the conclusions that necessarily follow from it are equally certain. It is therefore certain that all space known to us is not empty, but is filled with a substance, the ether, which can be thrown into vibration. But whereas our knowledge of the geometrical relations of the processes in this substance is clear and definite, our conceptions of the physical nature of these processes is vague, and the assumptions made as to the properties of the substance itself are not altogether consistent. At first, following the analogy of sound, waves of light were freely regarded as elastic waves, and treated as such. But elastic waves in fluids are only known in the form of longitudinal waves. Transversal elastic waves in fluids are unknown. They are not even possible; they contradict the nature of the fluid state. Hence men were forced to assert that the ether which fills space behaves like a solid body. But when they considered and tried to explain the unhindered course of the stars in the heavens, they found themselves forced to admit that the ether behaves like a perfect fluid. These two statements together land us in a painful and unintelligible contradiction, which disfigures the otherwise beautiful development of optics. Instead of trying to

conceal this defect let us turn to electricity; in investigating it we may perhaps make some progress towards removing the difficulty.

What, then, is electricity? This is at once an important and a difficult question. It interests the lay as well as the scientific mind. Most people who ask it never doubt about the existence of electricity. They expect a description of it —an enumeration of the peculiarities and powers of this wonderful thing. To the scientific mind the question rather presents itself in the form—Is there such a thing as electricity? Cannot electrical phenomena be traced back, like all others, to the properties of the ether and of ponderable matter? We are far from being able to answer this question definitely in the affirmative.   (Hertz 1889, 314–15)

To explain polarization effects, Young and Fresnel claimed that light must represent a purely transverse wave traveling within a medium called the ether. To appreciate what this classification means, consider two ways in which a wave might propagate along a 1-dimensional substrate such as a string. The first of these—the "longitudinal" waves— constitute the sound waves that arise when we strike a stiff bar at one end with a hammer. These waves represent packets of increased compression that distort in

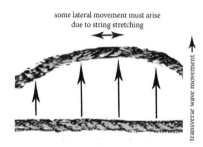

some lateral movement must arise due to string stretching

transverse wave movement

the same x-direction in which the wave itself heads. In contrast, *transverse waves* distort perpendicularly to the direction of advancement (viz., in the y- or z-direction). These pulses can be observed traveling along a flexible piano wire that has been pushed in a lateral direction by the instrument's hammers. The troubles arise when it is realized that in case of light, the posited wave movements must be *purely transverse*, unaccompanied by any longitudinal wiggling at all. But can any plausible medium truly exhibit transverse movements without simultaneously requiring a certain degree of accompanying distortion in a lateral direction? A glance at our string displays the problem: each little stretch of wire in a transverse pulse must elongate a wee bit in the x-direction to accommodate its y-direction bending. As this happens, tiny compression waves will begin to propagate along the string longitudinally, ruining the purely transverse character of the posited waves. When we first learn about strings in Physics 101, we tacitly ignore these x-direction complications on the grounds that their magnitudes will remain small enough to be ignored within the short term. By doing so, we can reach the simple uncoupled wave equation ($\partial^2 y/\partial t^2 = c\partial^2 y/\partial x^2$) familiar to every beginning student of the subject. But this convenient separation of wave types is merely approximative; over time energy will inevitably transfer between the string's longitudinal and transverse modes of movement in a complicated manner (Keller 1959).

Such an unavoidable loss in wave movement "purity" creates few problems for violin strings, but these crossovers present great difficulties with respect to light propagation within a material ether because their wave patterns are obliged to wiggle in a purely transverse manner over extremely long distances and intervals of time (Maxwell invites us to consider the unaltered spectral purity of the light that arrives from far away Arcturus). *Prima facie*, the hypothesized "ether" appears to demand material capacities that prove subtly incompatible with the tenets of classical continuum mechanics as they are conventionally understood today.

For reasons that we'll discuss in a moment, the odd form of mechanics that Hertz himself favors is not subject to the same modeling limitations. As noted before, Hertz outlines three basic "images" (= starting points) in which classical mechanical practices might become potentially clarified and sketches some of the standard deficiencies of these rival formalizations which he made no attempt to rectify (while allowing that such routes might prove feasible).[1] However, he did *not* include among these "images" the "continuum mechanics" policies to which I just alluded, despite that fact that modern engineers generally regard it as the best encapsulation overall of classical technique (Hertz overlooked such an approach because of the subject's mathematical complexities, for reasons that I'll survey in this chapter's appendix). Modern commentators on Hertz frequently presume that these different treatments should be regarded as descriptively "more or less equivalent," but this assessment is not correct when the feasibility of a purely transverse ether is concerned, as Hertz himself recognizes:

All physicists agree that the problem of physics consists in tracing the phenomena of nature back to the simple laws of mechanics. But there is not the same agreement as to what these simple laws are. To most physicists they are simply Newton's laws of motion. But in reality these latter laws only obtain their inner significance and their physical meaning through the tacit assumption that the forces of which they speak are of a simple nature and possess simple properties. But we have here no certainty as to what is simple and permissible, and what is not: it is just here that we no longer find any general agreement. Hence there arise

---

[1] The first of these "images" is a mechanics relying upon both geometrical constraints and Newtonian forces of an action-at-a-distance character, whereas Hertz's favored "third image" only tolerates the constraint forces ("stay out of my territory") that arise from the direct surface contacts of extended rigid parts. His second "image" attempts to work directly from the conservation of energy using Hamilton's principle, allocating specific quantities of energy to a system's available coffers of potential and kinetic energy. Hertz's favored "third image" scheme operates in much the same manner but eliminates Hamilton's potential energy storage capacities and overcomes some additional deficiencies in the long-range character of the variations his scheme requires. I'll add a few more details in the appendix, but none of these technicalities should affect the tenor of our central discussion. However, the parties who rashly assume that "of course, the Hamiltonian format perfectly captures everything there is to classical mechanics" should attend more carefully to the difficulties that Hertz raises with respect to non-holonomic coordinates.

actual differences of opinion as to whether this or that assumption is in accordance with the usual system of mechanics, or not. It is in the treatment of new problems that we recognize the existence of such open questions as a real bar to progress. So, for example, it is premature to attempt to base the equations of motion of the ether upon the laws of mechanics until we have obtained a perfect agreement as to what is understood by this name. (Hertz 1894)

These concerns provide a nice illustration of a dilemma of *inductive warrant*: seemingly equivalent or near-equivalent formulations of classical mechanics can differ significantly with respect to their advice on certain crucial questions. Kuhn (1962) is widely celebrated for claiming that classical mechanics established a clear paradigm for the "normal science" conducted under its aegis, but Hertz's worries about the ether demonstrate that established "paradigms" of a Kuhnian character often prove inconclusively silent with respect to certain vital veins of required research. Hertz is not interested in founding a Kuhnian "revolution": he merely wants to know "which way should we turn when we think about light?"

Let me now explain why I've selected the odd phrase "inductive warrant" to capture these considerations. In scientific use, which doctrines get dubbed as "laws of nature" represents a complicated affair, sometimes reflective of accidental historical contingencies. But in other situations the term plays an important discriminatory role, of the sort with which we are presently concerned. Hertz wants to ascertain upon an absolutist basis whether purely transverse waves are compatible with established physical tenet or not. The fact that many materials obey the linear principle known as "Hooke's law" does not contribute to this assessment at all, for we do not expect all materials will obey constitutive equations of this restricted character. But we feel differently about the conservation of energy and the balance of linear momentum, for the "inductive warrant" they extend to general wave processes strikes us as far more secure. As we've already observed, if the slate of fundamental continuum principles is completed in the canonical manner of our modern textbooks, no ponderous material is capable of serving as an adequate ether. But these assessments are delicate, and I've further indicated that more radical resolutions of the difficulty are possible. It is conceivable that a mechanism-based approach to mechanics articulated in Hertz's favored "third image" manner might tolerate purely transverse movements once the account is extended to encompass continuously many degrees of freedom. I don't know the final answer here, because Hertz never advanced to a consideration of the "limits" that he vaguely mentions in alluding to continuous media (Hertz 1894, 47).

I stress the delicacy of these subtle "law"-based discriminations because analytic philosophers only worry about the gross distinctions between "laws of nature" and "accidental generalizations." But a more finely grained ranking of principles is wanted to resolve the vital discriminations that Hertz hopes to address with respect to mechanical viability. I have no developed account to offer on this score, except to point out that the "inductive warrants" we can extract from empirically verified "laws" may differ considerably from one another with respect to applicability and salience.

In the "what is electricity?" portion of his 1889 lecture, Hertz alludes to another difficulty that was widely discussed at the time. Maxwell's equations link the electromagnetic field to its presumed sources in an oddly incomplete way: static charges excite the field in one manner and currents in another. What doesn't a current simply represent a flow of charged electrons in a common direction? Well, that is what we believe today, thanks to the efforts of Henrick Lorenz, but such a dualist account (particles + fields) was not widely accepted or well developed at the time Hertz wrote. Several popular alternatives (the so-called vortex proposals of Kelvin and Lamour) instead identify "ponderable matter" with singularities within the ethereal flow, in which case the notion of an "electrified particle" will not constitute an identifiable "thing" apart from the ether. Understanding the behaviors of light inside a transparent medium such as glass will depend vitally on how this notorious "field versus particle" dichotomy is resolved.

Here's the flowery language in which Hertz clearly expresses his distrust of action-at-a-distance interaction and his interest in ether-based theories of "ponderable matter":

Not less attractive is the view when we look upwards towards the lofty peaks, the highest pinnacles of science. We are at once confronted with the question of direct actions-at-a-distance. Are there such? Of the many [types] in which we once believed there now remains but one—gravitation. Is this too a deception? The law according to which it acts makes us suspicious. In another direction looms the question of the nature of electricity. Viewed from this standpoint it is somewhat concealed behind the more definite question of the nature of electric and magnetic forces in space. Directly connected with these is the great problem of the nature and properties of the ether which fills space, of its structure, of its rest or motion, of its finite or infinite extent. More and more we feel that this is the all-important problem, and that the solution of it will not only reveal to us the nature of what used to be called imponderables, but also the nature of matter itself and of its most essential properties—weight and inertia. The quintessence of ancient systems of physical science is preserved for us in the assertion that all things have been fashioned out of fire and water. Just at present physics is more inclined to ask whether all things have not been fashioned out of the ether? These are the ultimate problems of physical science, the icy summits of its loftiest range. Shall we ever be permitted to set foot upon one of these summits? Will it be soon? Or have we long to wait? We know not: but we have found a starting-point for further attempts which is a stage higher than any used before. Here the path does not end abruptly in a rocky wall; the first steps that we can see form a gentle ascent, and amongst the rocks there are tracks leading upwards. There is no lack

of eager and practiced explorers: how can we feel otherwise than [be] hopeful of
the success of future attempts?   (Hertz 1889, 327)

Further concerns of inductive warrant affect the status of the conservation of
energy, originally articulated in the 1847 essay by Hertz's own teacher, Hermann
Helmholtz. At that time Helmholtz derived the principle within a framework of
point masses connected over distances by so-called central forces. This assumption allowed him to introduce a force-based notion of "potential energy"[2] which
supplies the ingredient that he adds to traditional kinetic energy to supply the total
energy that remains conserved within an isolated system. But these same central
force assumptions seem ill-suited to physical systems (such as the ether) that
transmit their effects through direct contact rather than action-at-a-distance. In
an 1891 tribute to Helmholtz's work, Hertz describes how Helmholtz subsequently searched for an alternative route to total energy conservation that relies
instead upon a variational principle of some type:

> It is not generally known that in his mature years Helmholtz has returned to the
> work of his youth and has still further developed it. The law of the conservation
> of energy, general though it is, nevertheless appears to be only one half of a still
> more comprehensive law.... In the case of purely mechanical systems it has long
> been known that every system, according to the conditions in which it is placed,
> arrives at its goal along the shortest path, in the shortest time, and with the least
> effort. This phenomenon has been regarded as the result of a designed wisdom:
> its general statement in the region of pure mechanics is known as the Principle of
> Least Action. To trace the phenomenon in its application to all forces, through
> the whole of nature, is the problem to which Helmholtz has devoted a part of the
> last decade. As yet the significance of these researches is not thoroughly under-
> stood. An investigator of this stamp treads a lonely path: years pass before even a
> single disciple is able to follow in his steps.   (Hertz 1891, 338)

As it happens, Hertz adopted a somewhat different strategy to achieve this same
end, which I will outline in this chapter's appendix.

(iii)

Such are the significant challenges of ascertaining inductive warrant that confronted Hertz, but tightly aligning these large-scale concerns with the picayune

---

[2] In particular, forces of these specific types can be derived from a "potential" in the manner of
Lagrange and Green, which had been previously viewed as a device for evading the inconveniencies of a
vectoral notion (force) in terms of a scalar (potential).

topics emphasized in the preface to *The Principles of Mechanics* is often difficult. The basic reason for this lacuna stems from a mathematical handicap that affected almost all mechanical research in this era: no one fully understood how the mathematics suitable for a truly flexible body (a so-called "continuum physics object") should be laid out (this task required fifty-odd years of research after Hertz's time before adequate foundational proposals were developed).

As we've seen, the hypothetical ether appeared as if it must exemplify some elusive mixture of the behaviors exemplified by continuous solids or continuous fluids. But no one knew how to specify either of these behaviors in a direct way and instead retreated to considering finite degrees of freedom systems coupled with some hazy handwaving with respect to eventual "limits." In particular, textbook expositions commonly formulate their governing tenets in terms that only make clear sense with respect to finite swarms of mass points. Newton's First Law of Motion supplies an illustration in point:

> Every body persists in its state of being at rest or of moving uniformly straight forward, except insofar as it is compelled to change its state by an impressed force.

But if such a "body" is outfitted with any extension at all, at which of its constituent points are we supposed to ascertain this constant velocity? (Suppose that our "body" is a spinning ball.) Likewise, when an outside force impresses itself upon our "body," upon which locale will this force be impressed? (Again our ball will respond differently according to the "English" with which the force is applied.) But if we restrict our "bodies" to 0-dimensional points, these ambiguities become immediately resolved.

Now it is entirely apparent that Newton's intended universe did not consist in mass points alone; indeed, he appears to regard the latter as merely providing convenient mathematical reductions with respect to more extended entities. By Hertz's time, few physicists believed that the universe was actually composed of points or small rigid bodies, although intermediate earlier figures had certainly done so.[3] But if we judge these classical authors by the mathematical letter of their initial "foundational" pronouncements, we will likely assign them "ontologies" comprised entirely of point masses.[4] Significant misunderstandings of early modern authors frequently trace to this rash assumption.

In any case, a standard Victorian textbook will begin with a finite array of discrete point particles but later smush them together until their boundaries

---

[3] E.g., Descartes with respect to rigid bodies and Boscovich and some of the French atomists with respect to point masses.

[4] Such ambiguities of application are apparent in the very examples that Newton appends to his basic Three Laws. So much the worse for Quine's inadequate standards of "ontological commitment," I would say.

miraculously disappear and a smooth continuum is left behind. Here's the form in which this tactic appears in Hertz:

> In what follows we shall always treat a finite system as consisting of a finite number of finite material points. But as we assign no upper limit to their number, and no lower limit to their mass, our general statements will also include as a special case that in which the system contains an infinite number of infinitely small material points. We need not enter into the details required for the analytical treatment of this case.   (Hertz 1894, 46–7)

In truth, he would have been better advised to have "entered into those details," for no competent mathematician today could possibly regard his unconstrained appeals to "limits" as satisfactory. Inspired by Hertz but cognizant of these dangers, David Hilbert arranged the task of axiomatizing mechanics and their "limits" upon his famous 1899 list of twenty-three problems that mathematicians should address in the century ahead (Hilbert 1902, 454–5).[5] Hilbert's prompting suggestion eventually led to the more satisfactory analysis that classical continuum physics eventually received in the 1950s (further refinements and alterations continue to this day) (Maugin 2017).

In this book, I will skirt the intricacies of continuum physics as much as possible, for we will be able to appreciate the merits of non-axiomatic forms of conceptual clarification without delving into the arcane issues required (some of which will be canvassed in the appendix). But there is at least one lingering theme that derives from these concerns that we have already noted: the methodological presumption that "science always idealizes," a slogan that frequently serves as a pitiable excuse for brushing aside outright descriptive inadequacy. Historically, this theme drew much of its original sustenance from the presumption that classical physics must set aside its favored entities (viz., continuous flexible bodies) in favor of simpler "idealized" constructs (viz., point masses) upon whose behaviors applied mathematics can more easily gain a coherent conceptual grip. From a modern perspective, such appeals were premature and merely illustrate the popular ploy of invoking "philosophical" doctrines as a crutch to bridge deductive gaps when honest mathematics is properly needed. Certainly, the dodgy "limits" to which such tactics generally appeal have been replaced by the careful study of "homogenization" relationships that we shall briefly discuss in Chapter 6. In any case, unconstrained appeals to "idealization" inherently merit

---

[5] Hilbert dealt with both integral equations and early notions of "weak solutions" in his own work; both notions become important within the subsequent twentieth-century research in continuum mechanics. Much of this was initiated by his student Georg Hamel and later refined by the school of Clifford Truesdell and Walter Noll (Hamel 1921; Truesdell 1977).

closer critical scrutiny than they generally receive at the hands of philosophers (Wilson 2017, 130–5).

We here confront a strange development that has seriously handicapped philosophy's ability to profit from these old debates. Contemporary methodological discussion is strongly dominated by frequent appeals to the "worlds of classical mechanics" or "the classical mechanics picture" as if these phrases possess pellucid meanings that everyone recognizes. It is common, for example, to find these "worlds" offhandedly characterized as implementing "the classical 'billiard balls' picture of mechanical behavior," despite the fact that none of this phraseology can withstand any degree of critical scrutiny. All of these formal presumptions depend heavily upon a certain *mythology of the successfully axiomatized classical mechanics* which developed in the logical positivist period and has been perpetuated ever since, serving as a deceptive beacon whereby "formal philosophers" have thereafter attempted to guide their frigates. The basic assumptions that comprise this formal myth are: (1) classical mechanical doctrine, including all of its relevant laws, have been successfully and completely reduced to axiomatic form; (2) this axiomatization rests upon an ontology of point masses that non-experts can easily understand; (3) the "possible worlds" of classical thinking can be conveniently identified with the "models" in the logician's sense that these governing axioms accept; (4) within this framework, all physical explanation reduces to simple forms of what I'll call "initial condition based models"; (5) the scheme's use of point masses provides a useful illustration of the maxim that "successful science always idealizes."

These five presumptions often serve as the gold standard of successful clarification that work in "formal philosophy" has generally sought to emulate within an array of other applications. But all of these five claims are significantly mistaken, and in the next two chapters we shall utilize Hertz's perceptive insights with respect to assumptions (1) and (2). Although I will try to keep our discussion as simple as possible, the "small metaphysics" confusions that troubled Hertz are rather subtle in the "painful contradictions" they evince, so the nitty-gritty mechanical concerns that we will examine in the next chapter demand a certain degree of attention to nuance. The conceptual seeds from which grander philosophy later grows often reside in pesky small details such as these.

As it happens, Hertz's true purposes in composing *Principles* have frequently been wrongly identified afterwards due to a faulty confidence in the five doctrines just outlined. Because of these distortive presumption, later commentators often excuse Hertz of his apparent "misunderstandings": "He couldn't have been so stupid as to ignore these basic facts, so of course his objectives must have been different than what he actually claims." But the interpretative mistakes lie with the commentators, not with Hertz (who is generally quite careful in his claims). We shall examine several examples of these dubious "charitable interpretations" in the next chapter.

(iv)

But why has this axiomatic mythology so firmly taken root in the conceptual arsenal of contemporary philosophy? The answer traces to the fact that, beginning in the early twentieth century, a cloud of collective amnesia swept over physics and philosophy that degraded the substantial issues with which Hertz and his contemporaries struggled into simplistic "methodological lessons from the past" that present-day philosophers still employ in justifying their own investigative excursions. The excuse "I am merely engaging in effective idealization, as all good scientists do" (tenet 5) represents but one of these dubious "lessons," its roots within the mathematics of continua having become entirely forgotten. Distortions (3) and (4) reside in the fact that the same philosophers generally presume that they completely understand "the possible worlds" contemplated by the Victorian physicists, having forgotten the true controversies of the era.

A second source of classical mechanics forgetfulness stems from the fact that the active attention of most physicists abruptly shifted to the novel concerns raised by relativity and (particularly) quantum mechanics in the early twentieth century. Due to a fortuitous formal analogy, the earliest quantum models were constructed by "quantizing"[6] point mass models constructed according to the point mass recipe outlined below. Textbooks written for physicist (in contrast to practical engineers) strongly emphasize these classical point mass models while most of the more complicated continuum physics constructions have disappeared from their pages. This shift in emphasis in no way reflects the descriptive superiority of the point mass models; it is merely the outcome of the pedagogical needs of quantum physicists in training. But philosophers who derive their conceptions of "classical physics" from the materials they encounter within their Physics 101 classes often develop inadequate impressions of what the subject is really like as a result (they'd have done better if they had enrolled in Mechanical Engineering 101 instead). It is truly astonishing how swiftly this doctrinal amnesia occurred. Thus we find the astrophysicist Arthur Eddington confidently pronouncing in 1928:

> The Victorian physicist felt that he knew just what he was talking about when he used such terms as matter and atoms. Atoms were tiny billiard balls, a crisp statement that was supposed to tell you all about their nature in a way that could never be achieved for transcendental things like consciousness, or beauty or humor. (Eddington 1928, 259)

But Eddington was evidently familiar with Karl Pearson's 1892 *The Grammar of Science*, for he liberally borrows metaphors from it. In the appendix, we shall

---

[6] That is, by replacing ODE terms with Schrödinger equation operators.

briefly survey the strange opinions that Pearson actually maintained with respect to atoms. I can't comprehend how a reader of that book could seriously maintain that "the Victorians knew just what they were talking about." Quantum mechanics, after all, had only been in full bloom for about three years when Eddington wrote. How freely we attribute simple-mindedness to our elders, despite the fact that they may have been considerably more sophisticated in their own thinking than we are!

As we shall see in a moment, this modern point mass surrogate for "classical mechanics" is remarkably simple in its formal contours, and this very simplicity has greatly encouraged the reductive schematism that unfortunately characterizes a large part of current philosophical thinking with respect to "theories." A central tenet of this book claims that successful scientific explanation within classical mechanics relies upon a wide array of "reasoning architectures," of which the point mass formalism comprises a rather misleading specimen. I won't attempt to clarify what I intend by a "reasoning architecture" at this point but will wait until we have some clear contrastive specimens before us (in Chapter 5). But let me briskly outline the basic "architecture" of the point mass scheme I have in mind (which I'll call "Euler's recipe" in the sequel,[7] at the price of a certain degree of labeling injustice). Readers unfamiliar with differential equations may skip these details.

The task of supplying a suitable axiomatization for the expected point mass behaviors proves relatively easy, at least initially. Select a target system $\mathfrak{S}$ to model in a point mass mode. Against the backdrop of standard Euclidean 3-space, enumerate the finite number of masses included in $\mathfrak{S}$, as $1, 2, \ldots, n$. For each $i \, \varepsilon \, \mathfrak{S}$, write down the following collection of n ordinary differential equations employing vector notation:

$$m_i d^2 \mathbf{q}_i(t) dt^2 = \Sigma \mathbf{f}_j(i, t).$$

In these formulas, m_i represents the mass of the particle numbered as "i," $\mathbf{q}_i(t)$ represents its vector location at time t and the various $\mathbf{f}_j(j,t)$ supply the strengths of specific forces applicable to particle i that trace to a source located at the particle j (for $i \neq j$). These action-at-a-distance equations represent reasonable precisifications of Newton's Second Law (although not in a manner that the historical Newton would have necessarily embraced). Newton's Third Law places rather severe restrictions on their interdependencies within these special force laws (viz., $\mathbf{f}_i^X(j,t)$ must equal $-\mathbf{f}_j^X(i,t)$ for every special force type X).[8]

---

[7] I have labeled this familiar ODE framework as "Euler's recipe" because Euler was the first mathematician to clearly identify Newton's "$\mathbf{F} = ma$" as a formula upon which a fuller descriptive scheme can be formally centered. But the format doesn't render justice to Euler's actual attitudes with respect to "mechanical foundations."

[8] These supplementary stipulations are often called "the strong form of the Third Law" and are generally invoked in deriving the conservation of energy within our Eulerian recipe. As we'll see in Chapter 4, these strong law provisos are the key source of the tensions with constraint-type forces at the

But these equations merely lay down a *scaffolding* upon which we might potentially construct an adequately closed set of equations.[9] To fill out this collection, we must plainly specify laws for how the otherwise unknown "special force function" terms $f_j(i,t)$ appearing in the "$\sum f_j (i, t)$" term should act. The basic prototype for a "special force law" of this ilk is Newton's own law of gravitation: $f_i^G(j,t) = (G. \, m_i. \, m_j)/|q_i(t) - q_j(t)|^2$ where G is the universal gravitational content. But further allied principles will be required to govern whatever additional forces enter the scene, including those responsible for the cohesions and repulsions of ordinary matter. And here the Euler's recipe tradition turns suspiciously silent. To be sure, there are various semi-plausible principles that closely follow the gravitational paradigm, but they are plainly not universally applicable.[10]

So we appear to have made a good start in developing a proper axiomatics for point mass classical mechanics. But then the trail peters out. We still lack the requisite "special force laws" required to carry a target system forward in time from initial conditions (under the presumption that interactions other than gravitation are required). But where are they? I have elsewhere dubbed this palpable lack of equational closure as "the mystery of the missing physics" (Wilson 2009). In truth, this failure to fully equip our ODE recipe with an adequate set of special force laws stems from the fact that field and quantum mechanical behaviors strongly intercede whenever matter acts within close quarters, and nature offers insufficient guidance in how such gaps should be filled with acceptably "classical" provisos. In the circumstances, dogmatically filling in the lacuna in order to possess a completed axiomatics seems a bootless enterprise. In real-life practice, classical physicists work around these missing ingredients by departing from the Euler's recipe framework and appealing to other descriptive resources to flesh out a well-set mathematical problem. A commonly employed "work around" tactic of this type appeals to *constraints*: higher-scale information that restricts the movements of the point masses to fixed surfaces. A standard exemplar of such an appeal can be found in the perplexing "beads on a wire" that are

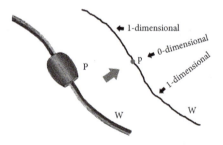

center of Hertz's complaints. I might also add that alternative readings of Newton's strangely worded principle ("action equals reaction") create other confusions as well, but I shall try to avoid these as best I can.

[9] I.e., a set of stipulations that might prove uniquely solvable (at least most of the time) with respect to an arbitrary set of initial conditions (supplementary "data" that provides $q_i(t_0)$ and $dq_i(t_0)/dt$ information for each particle i∈S at some selected time t_0).

[10] As example of the former, Coulomb's law for static electrical attraction: $f_i^C(j,t) = (k_e. \, c_i. \, c_j)/ |q_i(t) - q_j(t)|^2$ (where the c_j can carry negative values) can be cited. As an exemplar of the second type, the 1/6 power law often employed for molecular repulsion upon close approach.

offhandedly introduced within the opening pages of every elementary physics primer. In these treatments the "bead" is modeled as a point particle p confined to the surface of a rigid wire W, which is mathematically characterized as a continuous, one-dimensional geometrical curve. But shouldn't W itself be properly modeled as a swarm of point masses if we are to remain loyal to our Euler's recipe provisions? And won't additional special force laws be needed to confine p's movements within a close proximity to W without generating a large degree of frictional resistance in the process? Our instructor in Physics 101 might cheerfully concede, "Of course" but evince little interest in filling in the gaps. He or she may then direct you to a more advanced class that relies upon a significant amount of quantum mechanics that sidesteps your classical Euler's recipe worries.[11]

In point of fact, these casual appeals to constraints adjust the underlying "explanatory architecture" in significant ways, and Hertz's worries about the proper "inductive warrant" of classical mechanics with respect to the ether ultimately trace to this little shift in underlying strategic support. I will document how this happens in Chapter 5. For the time being, let us simply stipulate that the co-mingling of point masses with constraints represents a *hybrid descriptive scheme*, in which distinct mathematical ingredients appear to cooperate with one another successfully. Alternative formalisms of the types developed by Lagrange and Hamilton were especially designed to facilitate these blendings of ingredients.

A strong reason for pursuing descriptive policies of this mixed character stems from a commendable interest in heightening descriptive *reliability*: if we *already know* that the bead p will remain in the proximity of the wire W, why engage in a hypothetical point mass modeling with respect to its alleged supportive causes, especially when the true story of these attractions involves quantum mechanical behaviors that cannot be easily replicated by any classical "special force law" mechanisms? Stated another way, quantum "laws of nature" are undoubtedly responsible for the cohesions and stiff rigidities of macroscopic wires, but these principles cannot be plausibly recast to satisfy the "classical" demands of our Euler's recipe axiomatics. As we already observed, Maxwell and Kelvin frequently extolled the descriptive advantages of appealing to constraints in Lagrange's hybrid manner because such tactics allow us to skirt lower-scale speculations of

---

[11] Michael Spivak satirically describes how one of these force-avoiding workarounds surreptitiously enters into a standard textbook's discussion of billiard ball collisions:

> Conservation of momentum gives us only one equation.... Elementary physics textbooks need to provide problems that have answers, of course, so, in the manner of a host nonchalantly introducing a celebrity at a party, they will often unobtrusively insert a new definition: a collision is called "completely elastic", if we also have conservation of kinetic energy.... Once we have made a definition, it's possible to pose all sorts of simple problems about collisions that are "completely elastic", whatever that might mean.   (Spivak 2010, 94–5)

Spivak's volume is one of the best sources available for exposing the little inconsistencies that riddle standard textbook practice.

an unsubstantiated nature. In doing so, we greatly reduce the risk of unwanted descriptive errors.[12] The multiscalar schemes of Chapter 5 will elevate these commonsensical methodological observations into certifiable "architectures."

(v)

Before we close the book upon Euler's recipe, let me hasten to observe that the devoted study of point mass behaviors represents a very important branch of mathematical study that is commonly labeled as "celestial mechanics" or simply "classical mechanics." Lord Kelvin characterizes these dedicated studies as:

> the never-ending interest in the definite mathematical problem of the equilibrium of motion of a set of points endowed with inertia and acting upon one another with any given force.   (Thomson 1904, 123)

But we go badly astray if we identify such studies with the full range of "classical mechanics" historically conceived.[13]

It is exactly in opposition to a point mass perspective that Hertz embraces a distinct resolution that has not been adequately appreciated by many of his readers. As we shall see in more detail in Chapter 4, when Hertz writes of mechanical doctrine as containing "more relations than can be completely reconciled amongst themselves" (Hertz 1894, 7), he is considering the hybrid mixing of point masses with constraints that we have just discussed. As this chapter's appendix further details, Hertz's central project in *Principles* is precisely one of reducing this excessive doctrinal baggage by purging point masses and their conjoining action-at-a-distance forces from his foundational doctrines in favor of the constraint relationships alone. In doing so, the scheme embraces an organizational ontology that

revolute   prismatic   screw   rolling ball

---

[12] One of the chief characteristics that distinguishes real science from pseudo-science traces to the fact that the former generally reinforces its speculations with strong fibers of reliable doctrine. Bypassing ungrounded speculations about molecular constitution whenever possible represents an important reliability-enhancing policy of this general character. Conventional thinking about "scientific theory" often overlooks these prudent safeguards (this general theme of "tempering speculation with reliability" will become central in Chapter 5).

[13] The scheme's stark mathematical simplicity continues to shape philosophical thinking about science in unfortunate ways even to this day. Point mass celestial mechanics is easy to grasp conceptually and serves as a central paradigm that sustains the "theory T thinking" of which I complain throughout this book. I have somewhere seen these tactical distortions characterized as "looking at PDEs through ODE eyes." But when I try to locate my source via internet search, I only find my own writings! For fuller fulminations on these themes, see Wilson (2017).

operates with connected mechanical ensembles that were labeled as (iii) in Chapter 2), viz., "continuously contacting rigid bodies in the fashion of a gear chain or clockwork."[14] The standard label for this kind of mechanical construction is "linkage," indicating a collection of rigid parts (the "links") that are constantly maintained in close but adjustable contact with one another by purely mechanical interconnections such as a hinges, pins, interlocked gear wheels, balls rolling upon surfaces, cams and followers, etc. The illustration displays four of these basic constraint assemblies. Such relationships can be captured in either algebraic (most of them) or differential equation (the rolling ball) terms as so-called "constraint equations," usually without reference to time. Hertz's proposed corrective expunges all of the special force laws utilized within our Euler's recipe scheme in favor of these constraint relationships. In doing so, the "contact forces" that bind these contraptions together remain intact, but all of the familiar Newtonian action-at-a-distance forces (including gravitation) are banned.[15] Unfortunately, Hertz's favored terminology for a linkage is "a system of material points," phraseology that has led a fair number of modern commentators astray, who fancy that Hertz seeks a "metaphysically purged" reorganization of our Eulerian point mass recipe. But the revisionary doctrines that Hertz strives to rigorously axiomatize are far more radically conceived than that and become far stranger in their operational details than we might naïvely anticipate. As Arnold Sommerfeld pithily comments:

[H]is is an interesting and stimulating idea, carried out with great logic; because of the complicated replacement of forces by connections it has, however, borne little fruit.   (Sommerfeld 1943, 214).[16]

---

[14] It is important to distinguish this mechanism-centered approach to rigid body behavior from the more familiar billiard ball modelings discussed in Chapter 5. The latter admit of instantaneous impacts which Hertz sternly bans from his universes:

All connections of a system which are not embraced within the limits of our mechanics, indicate in one sense or another a discontinuous succession of its possible motions. But as a matter of fact, it is an experience of the most general kind that nature exhibits continuity in infinitesimals everywhere and in every sense: an experience which has crystallized into firm conviction in the old proposition—*Natura non facit saltus*.   (Hertz 1894, 37)

In contrast, Hertz's scheme is what mathematicians call "smooth": his foundational rigid bodies can only slide or roll across one another's surfaces with well-defined velocities and can never abruptly rebound or fracture. His reasons are much the same as those discussed in relation to Leibniz and du Châtelet in Chapter 6.

[15] This policy is frequently described as "doing without forces," but the characterization "retaining only constraint forces" strikes me as less misleading. We might recall that providing a plausible mechanism for stable material cohesion represents a significant challenge within a Euler's recipe setting.

[16] Insofar as I can determine, most of the physicists close to Hertz's time appear to have grasped his foundational intentions clearly enough (Sommerfeld's exposition is admirable), and it is only the "logical empiricists" of Carnap's era who misconstrued the project as one of eliminating an unwanted "metaphysical" conceptual intrusion, despite Hertz's explicit declarations to the contrary (cited later in

For the central purposes of this book, we needn't delve into these rather arcane considerations, but I've tried to epitomize how they unravel in the appendix. I will only observe that I generally prefer the modern term "constraint" to the "connections" favored by Hertz and Sommerfeld.

I again stress that Hertz, like most of the other physicists of his era, is ultimately interested in the alternative form of *continuum physics* that we might obtain once we subject his finite arrays of linkages to infinitary "limits." As I've already indicated, no one in Hertz's time knew how to develop these "limits" with sufficient care (which required the later attention of David Hilbert and his school). When the proper mathematical tools become available, a wider dominion of generalized continua can be reached whose behaviors depart in significant ways from the conventional continua that do not tolerate purely transverse waves. The best known of these novel forms of "generalized continua" are the micropolar models that were developed by the Cosserat brothers in 1909. These media incorporate rod-like infinitesimals that can respond selectively to induced torques (fields of Cosserat type are commonly employed to model elastic materials that are sensitive to electromagnetic fields such as liquid crystals) (Vardoulakis 2019). By hoping for a "generalized continuum" limit based upon linkage-like infinitesimal elements, Hertz could plausibly hope that a "transverse wave only" ether might be placed upon a sounder physical basis. Indeed, Cosserat mediums can transmit waves of types that are not possible within media of a more conventional stripe. I don't know whether Hertz's goals can be legitimately achieved in this hypothetical fashion or not. Fortunately, this no longer represents a vital foundational project within ongoing physics.[17]

Here's a simple way of appreciating the dialectics currently at issue. Standard mechanical practice employs a large number of hybrid ingredients and deciding

---

this chapter). Later purported axiomatizations such as Simon (1954) pursue this latter (to my mind utterly misguided) project. These fictive ambitions serve centrally as motivational pretexts in van Fraassen (1980) and similar writings.

[17] Hertz wished to determine whether an ether suitable for electromagnetism could be realized within a ponderous medium. In our modern era, this task of matter/field alignment has essentially reversed itself. For ourselves, the electromagnetic field represents an unproblematic entity in its own right, whereas the task of inserting classical matter consistently into its confines becomes problematic. As a consequence, Hertz's original worries with respect to transverse waves have changed their complexion considerably. A few years after Dedekind, Poincaré captured this adjustment in attitude as follows:

> One is justified in asking if we are not on the eve of just such a revolution or one even more important. Matter seems on the point of losing its mass, its solidest attribute, and resolving itself into electrons. Mechanics must then give place to a broader conception which will explain it, but which it will not explain. So it was in vain the attempt was made in England to construct the ether by material models, or in France to apply to it the laws of dynamic. The ether it is, the unknown, which explains matter, the known; matter is incapable of explaining the ether.
>
> (Poincaré 1905, xiii)

which one qualifies as properly "foundational" can resemble the impossible task of identifying the lowest landing within one of those topologically twisted staircases for which the artist M. C. Escher was renowned.[18] Euler's recipe begins in the point mass corner of our platform and attempts to reach continuum physics by taking appropriate "limits." But Hertz contemplates starting instead at the far "rigid bodies" landing and reaching a wider class of continua by an unspecified collection of "limits." But we can't defini-

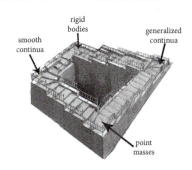

tively resolve our worries with respect to a transverse-wave-only ether until we decide which landing in our staircase is the lowest. Our Escher-like portrait of the conceptual situation indicates how the problem of identifying the "inductive support" provided by an otherwise successful descriptive practice can prove maddeningly elusive.

(vi)

With respect to Hertz's intentions, we should observe that Maxwell's celebrated "dynamical model" of the ether suggests the promise of such an approach, for when the little idle wheels corresponding to an electric current speed up or slow down in these arrangements, an impulse within the "magnetic" hexagons will move away from the current in an almost perfectly transverse pattern. Of course, neither Hertz not Maxwell himself fancied that such an implausible arrangement reflects electrical reality, but the construction suggests that the greater allowances of generalized continuum think-

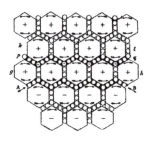

ing might lead to a more satisfactory basis for the purely transverse waves of electromagnetism. As already noted, Hertz's cavalier attitudes with respect to "limits" would have prevented him from pursuing such a program even if he had lived, and the   Cosserat brothers' own proposals were not seriously advanced until the 1960s.

---

[18] Our Escher figure exemplifies the twisted topology of a Riemann surface, which represents the central metaphor for "foundational looping" that I utilize in Wilson (2006, 312–19).

It is worth pausing to address Pierre Duhem's often-cited complaints that his British physicist contemporaries rely excessively upon the "visualizable models" found in both Maxwell and Kelvin:

> [Consider] the ample but weak mind of the English physicist. Theory is for him neither an explanation nor a rational classification of physical laws, but a model of these laws, a model not built for the satisfying of reason but for the pleasure of the imagination, Hence, it escapes the domination of logic. It is the English physicist's pleasure to construct one model to represent one group of laws, and another quite different model to represent another group of laws, notwithstanding the fact that certain laws might be common to the two groups.... Thus, in English theories we find those disparities, those incoherencies, those contradictions which we are driven to judge severely because we seek a rational system where the author has sought to give us only a work of imagination.
>
> (Duhem 1914, 80–1)

A tincture of justice may reside within this familiar, jingoistic caricature (which we shall briefly discuss in Chapter 6 in conjunction with Ernst Mach's milder complaints with respect to psychological bias). However, these British authorities should also be favorably regarded as practicing an entirely reasonable agnosticism with respect to mechanical foundations. As we've just seen, Maxwell's otherwise implausible model for ether suggests that Cosserat-like resolutions of the medium's difficulties should not be ruled out of bounds by any premature reliance upon a fixed "rational system" of a Duhemian cast.

In our own discussion, we needn't explore the rather peculiar ways in which Hertz eventually develops his machine-based mechanics, for they will not prove material to the central methodological morals I hope to draw (for completeness' sake, I will outline his chief tactics in the appendix below). I instead wish to emphasize Hertz's positive philosophical contributions in two central ways: (i) his skill in isolating an otherwise unnoticed tension at the core of established mechanical doctrine and (ii) his general portrait of how these conceptual disharmonies came to secretly shelter within the folds of a descriptive practice that had amply established its empirical credentials many times over.

Although Hertz's machine-focused orientation was accurately recognized by the physicists of his time, these same ingredients within his thinking are often absent in the commentary of more recent years (Baird, Hughes, and Nordmann 1998). Some of these misunderstandings stem from the fact that Hertz inadequately stresses the pathway between generalized continua and the finite systems that are directly described within the *Principles* (Hertz is notoriously terse in registering his motivations, especially when they partake of a speculative character). His persistent employment of "connection between material points" rather than the more familiar "constraint equation" generates further interpretational difficulties as well. And quite significant misunderstandings trace to the physics community amnesia described above, leading later authors to ascribe motivations to Hertz that represent the polar contraries of what he actually maintains. Here is a typical specimen:

> For Hertz it was the "metaphysical" nature of force and the puzzle over the necessary or contingent, a priori or empirical status [of $F = ma$] that made it especially desirable to rid the theory of that notion.   (Sklar 2013, 157)

But Hertz expressly states that he is not concerned in the least with the allegedly "occult" or "metaphysical" demerits of action-at-a-distance forces but merely with the fact that their prescribed properties have not been systematically or consistently specified:

> Weighty evidence seems to be furnished by the statements which one hears with wearisome frequency, that the nature of force is still a mystery, that one of the chief problems of physics is the investigation of the nature of force, and so on. In the same way electricians are continually attacked as to the nature of electricity. Now, why is it that people never in this way ask what is the nature of gold, or what is the nature of velocity? Is the nature of gold better known to us than that of electricity, or the nature of velocity better than that of force? Can we by our conceptions, by our words, completely represent the nature of any thing? Certainly not. I fancy the difference must lie in this. With the terms "velocity" and "gold" we connect a large number of relations to other terms; and between all these relations we find no contradictions which offend us. We are therefore satisfied and ask no further questions. But we have accumulated around the terms "force" and "electricity" more relations than can be, completely reconciled amongst themselves. We have an obscure feeling of this and want to have things cleared up. Our confused wish finds expression in the confused question as to the nature of force and electricity. But the answer which we want is not really an answer to this question. It is not by finding out more and fresh relations and connections that it can be answered; but by removing the contradictions existing between those already known, and thus perhaps by reducing their number. When these painful contradictions are removed, the question as to the nature of force will not have been answered; but our minds, no longer vexed, will cease to ask illegitimate questions.   (Hertz 1894, 7)

Note that he characterizes the desire for traditional conceptual clarification as a "confused wish" and maintains instead that the crucial analytic task is that of slimming mechanics' assembly of conflicting doctrines into a smaller, self-consistent set. He expressly allows that a similar winnowing might also work for a system of point masses interconnected by Newtonian forces, although he doubts that such a project can be easily accomplished. As such, Hertz's view of "conceptual analysis" is plainly grounded within our opening chapter's pragmatic concern with "articulating a doctrinal collection that works ably," rather than "clarifying murky ideas" in the manner of the "conceptual analysis" that Bertrand Russell favored. Under this second heading, I include the popular presumption that when conceptual tensions are found within a subject matter, a reconstructive analyst should scrutinize the internal contents of the terminologies at the core of her troubles until she espies the unnoticed conceptual ambiguities responsible for the conflicts. Russell (1914) provides a paradigm illustration of such an "analytic" approach to the tasks of philosophy, which I have characterized at some length as "the classical view of concepts" (Wilson 2006, 139–46).[19] In contrast, Hertz plainly believes (as I do myself) that when words are closely scrutinized in this microscopic fashion, there is frequently nothing to see, for the problems actually lie in how the wider fabric of the language has become stitched together, of which there may be no locally discernible sign. Insofar as the themes of this book go, the passage quoted effectively distills the vital philosophical lessons that we should extract from the *Principle*'s celebrated preface.[20]

Many of the prevailing misapprehensions with respect to Hertz's approach to conceptual repair derive from the common presumption that when Hertz writes of "systems of mass points," he intends to develop a revised Euler's recipe scheme cleansed of direct reference to "forces," regarded as "metaphysically" noxious ingredients. But this is clearly a misreading of the quoted passage. In fact, the Hertz's "mass points" merely represent convenient locations arrayed along extended rods and gears that we employ in locating their positions with a finite array of coordinate numbers. When he writes of a "system of material points," he is not referring to point masses in our Euler's recipe manner, but merely labeling a particular rigid part $\mathcal{R}$ by selecting convenient coordinates for $\mathcal{R}$'s location through appeal to (e.g.) its center of mass and axis of inertia (if $\mathcal{R}$ is allowed to rotate). Instead of writing that rod $\mathcal{R}$ maintains a rigid length, he appeals to a constraint relationship: any pair of convenient coordinate locations arranged

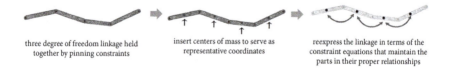

three degree of freedom linkage held together by pinning constraints   insert centers of mass to serve as representative coordinates   reexpress the linkage in terms of the constraint equations that maintain the parts in their proper relationships

---

[19] Quine (1969, 27) characterizes such views as "the myth of the mental museum."

[20] Similar sentiments can be found in Mach and Helmholtz. The passage cited served as the opening motto for Ludwig Wittgenstein's *Philosophical Investigations* over most of its developmental life. I happen to believe that Wittgenstein extracted morals from Hertz similar to my own, but I will not attempt to plunge deeply into the swamps of Wittgenstein exegesis. The latter's *Tractatus* displays other symptoms of Hertzian influence, in a manner I will sketch in a footnote to this chapter's appendix.

along $\mathfrak{R}$ must maintain a constant distance from one another. If one part rolls across the curved surface of another, a differential equation connection marks the relationship.

However, the full story of Hertz's approach to "material points" is rather complicated, because he adds a large number of supplementary machine parts (which he labels as "hidden masses") to his mechanical systems so that their behaviors will submit to a simple form of governing dynamical law. None of these rather peculiar details will concern us in the sequel, but their basic rudiments are reviewed in the attached appendix.

However, the misunderstandings just sketched of Hertz's methodological intentions plays a significant historical role in generating the dubious standards of "ersatz rigor" with which this book is primarily concerned. In particular, this misidentification of Hertz's project played an important inspirational role in fostering the strongly "anti-metaphysical" approach to conceptual clarification that we find in Carnap. "If we can only specify a self-consistent axiomatic scheme to completely monitor our uses of our 'theoretical' terminologies," the relevant line of thought goes, "we can rid ourselves once and for all of the bootless disputes about their metaphysical status." The fervent hope that progressive science could be shielded against unhelpful "metaphysical" intrusion through the "implicit definability" offered within the housing of a completely axiomatized theory favorably appealed to a large number of early twentieth-century thinkers who had grown weary of haggling with incomprehensible critics reeking of Hegelian pipe tobacco (for the comparable enthusiasms that developed in close parallel within pure mathematics, see Wilson (2021)). I believe that this positivist vision of science's liberation from the vagaries of murky "philosophical" cavils continues to animate the "ersatz rigorists" of the present day, even when some of them direct their efforts towards the axiomatic resuscitation of old-fashioned "metaphysical" themes.

These is also the reason why appealing to "completed theories T" does not merely represent a "convenient epistemological fiction" for Carnap and his apostles; it tacitly embraces a severely regimented account of so-called "theoretically supported meaning" of which we should remain profoundly suspicious (see Chapter 7 for more on this).

With respect to our own unfolding investigation, this "purging physics of unwanted metaphysical intrusion" understanding of Hertz's purposes suppresses the methodological observation that I view as his most important insight, viz., the notion that an evolving descriptive practice enlarges itself by continually adding supplementary reasoning policies that eventually clash against one another through no particular fault of any of the contributing parties. When these conflicts emerge into daylight, we needn't:

seek out more and fresh relations and connections but remove the contradictions existing between those already known.   (Hertz 1894, 7)

The ability to detect evolution-induced conceptual conflicts in this manner strikes me as a rare diagnostic skill, in which a canny investigator must follow the available clues without copious amounts of anticipatory guidance. In the next chapter, we will find Hertz demonstrating these praiseworthy analytic capacities with respect to the classical mechanical practices of his time. In Chapter 5, we'll then learn that these same conceptual tensions can be alleviated by tactics other than brute subtraction.

Those who cannot remember the past may not repeat it exactly, but they readily become susceptible to dubious lessons extracted therefrom. Our latest pair of chapters have revisited old scientific themes in possibly more detail than is strictly required for my didactic purposes. But it is undeniable that philosophy remains haunted by a significant number of ghosts from its classical mechanical past, and present-day discussions often represent unhelpful exercises in pushing these unreliable "lessons from science" around upon the same unyielding chess board, never revisiting the "small metaphysics" discrepancies from which these folkloristic strictures originally sprang. This is the methodological cage in which Carnap's fondness for "formal reconstruction" has unwittingly entrapped the rest of us, confining our deliberations to an excessively schematic plane, from whose airy heights every trace of the originally motivating conflicts has been thoroughly scrubbed. But the byways of real-life conceptual confusion do not permit such hygienic luxuries. We must remain prepared to scrutinize the humblest little clues carefully, for they frequently lead to unexpected conclusions.

APPENDIX

Historical Complexities

Although I frequently appeal to developmental considerations in this volume, their guiding purpose is not that of an adequate historical reckoning but merely to explicate how a certain species of logicized dogmatism came to adversely affect analytic philosophy over the course of the twentieth century. These developments, exemplified within the methodological thinking of Rudolf Carnap and W. V. Quine, eventually consolidated into the superficial standards of "ersatz rigor" of which this volume complains. In particular, this logical empiricist methodology unwisely ratified a schematic portrait of "how science explains" that actively discourages the deeper veins of critical appraisal that genuine progress in philosophy requires.

But our chronological overview must reach further into the late nineteenth century because the inspirational roots behind the theory T methodology develop from some well-intentioned efforts to correct some nagging inconsistencies within the standard modeling policies of classical mechanics tradition. As indicated in Chapter 1, many of the concrete remedies proposed by the great philosopher/physicists of this era have subsequently regarded inadequate, but the methodological dialectics they engendered should nonetheless be regarded as one of the great productive periods in the history of philosophical inquiry. This intellectual movement is characterized by an admirable willingness to reconsider prevailing doctrinal shibboleths in strikingly novel ways. For these same reasons, the qualities of intellectual reappraisal that we find in applied mathematics' recent corrections to Hertzian presumption also strike me as illustrative of the freer veins of conceptual inquiry that can benefit philosophical inquiry everywhere, even in subjects far removed from science per se.

As I have also indicated, a substantial portion of the unhappy path to ersatz rigor commences within Heinrich Hertz's entirely reasonable suggestions for how those "nagging inconsistencies" might be profitably addressed. Later generations of applied mathematicians have concluded that his suggested form of axiomatic remedy was not the medicine best suited to the malady, a vein of reconsideration with a great potential for teaching us a lot about how a regimented linguistic practice can productively correlate with physical circumstance. So the central plan of the book is simply to start with Hertz's perceptive detection of mechanical inconsistency and then explicate the rather different correction that the modern studies have suggested.

Unfortunately, there are all of those intervening years between then and now, during which academic philosophy ratified into schematic dogma many of the methodological suggestions that Hertz and his contemporaries had more innocently suggested. This history is considerably complicated by the fact that formidable mathematical obstacles stood in the path of a coherent development of the continuum physics that the practitioners of the late nineteenth century hoped to see, but had not yet developed the mathematical tools to do so. As often happens, philosophical appeals frequently rush in where coherent mathematics fails to tread. The upshot is a persisting methodological fog that obscures the central moral for which this book argues (viz., reasoning pattern within classical mechanics is better suited to a multiscalar architecture than to single-level axiomatics). The complications engendered by continuum physics are extraneous to this conclusion but must sometimes be acknowledged in order to understand the historical record. And these circumstances generate a cloudy mist that sometimes makes it hard to align the rather simple story of mechanical confusion coupled to its modern corrective that I want for my central narrative purposes. Hertz, in particular, wanders into many odd corners of applied technique that can be completely ignored insofar as my central narrative goes but which often make it hard to align my favored emphases with his actual words.

For these reasons, in this appendix I will attempt to outline the chief mathematical obstacles that stood in the path of the late nineteenth-century physicist as well as the rather peculiar gambits Hertz adopted in attempting to bring his axiomatic goals to fruition. None of this detail matters to my central narrative at all, except for the necessity of clearing away unwanted distractions. The single most unsatisfactory aspect of Hertz's discussion lies in the cavalier manner

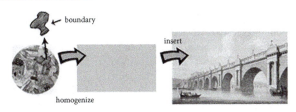

in which he expects to regiment the behavior of a continuous substance such as his ether by considering an infinite number of representative masses. The assumption is that somehow when one smushes together a large population of discrete particles, the limiting result will somehow emerge as a smoothly continuous system with any gaps or joins. Poof! But where did those earlier boundary joins go? I believe that I was not the only Physics 101 student who puzzled over these frequently invoked manipulations. Indeed, it turns out properly "homogenizing" a discrete population like this requires careful attention to the notion of "limit" (these issues will remerge briefly in Chapter 4). Such quandaries, in fact, appear to be the reason that David Hilbert placed the task of axiomatizing mechanics coherently as the sixth of his famous list of problems that the following century should address. He split his concerns into two modes that we might loosely label as "upward" and "downward looking." In the upward going mode, we want to determine in a reasonable way what lower-scale complexities will "look like" from a higher-scale point of view. We shall consider a number of questions of this type in Chapter 5 under the heading of "homogenizations. *Example*: suppose we consider a rock specimen that consists of tiny crystals jammed together along adjoining boundaries (such composites are frequently called "laminates"). How will these conglomerates behave on a higher size scale in which all of this finely grained disappears? Frequently, the correct answers are surprisingly simple: the rock may behave as a simple Hookean solid characterized by two basic parameters (its elastic constants). Extracting these mathematical conclusions rigorously generally demands the consideration of the limits that a representative population reaches as it becomes infinitely large (the classic illustration of this technique is the "central limit" construction whereby one extracts a simple bell-shaped curve from the individual activities of a large population of gamblers). And it further emerges that my Physics 101 worries were well founded; we must frequently pay close attention to those joining regions when we "homogenize" a laminate structure through a suitable limiting tactic. When we don't, we frequently fall victim to the many paradoxes that have bedeviled topics like the classical theory of structural beams over the years. The extensive amounts of functional analysis machinery that engineers currently devote to such concerns represent the direct product of investigations that were pioneered by Hilbert and his school.

But the "downward looking" suggestions that Hilbert also provides are especially salient to the project of axiomatizing classical mechanics because they propose that a physicist should directly deal with finite volumes of matter expressed as integrals, rather than attempting to assemble these units from smaller "particles" in the manner that Hertz and his contemporaries adopted. Hilbert's tactics utilize the "measure theory" techniques that have also proved critical in making sense of probability assertions with respect to continuous domains, such as Robin Hood's probabilities of hitting a target accurately. Here's the descriptive dilemma we confront. He is an excellent archer whereas Friar Tuck is not. But what is the probability that Robin's arrow will hit the exact point q on the target? Zero, because there are infinitely many equally likely nearby

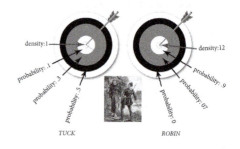

possibilities. What is the probability of Tuck hitting the same point? Again, zero for the same reason. Adding up all of the points on the target, we conclude that both sportsmen are equally and utterly inept. But this is ridiculous of course. How do we resolve the issue? *Answer*: we work downward from established probability assignments to regions of finite area *ℛ*. With respect to such a block, we can meaningly assert that "Robin's probability of landing within the region $ℛ_1$ is .9 whereas Tuck's is .15." Now consider a similar set of appraisals with respect to in a nested sequence of finite regions $\mathcal{R}_1$, $\mathcal{R}_2$, $\mathcal{R}_3$, ... that enclose the point q. If the respective probabilities divided by the appropriate areas tend to a final limit r, then we can meaningfully attribute the value r directly to the point q, not as a true probability (which remains 0) but as a so-called probability *density* (a "density" ρ is a function that supplies a genuine probability when ρ is integrated over the finite region $\mathcal{R}$). In this manner, a nested sequence of genuine probabilities will deposit (or "induce") a local density upon the individual point p. In doing so, we must verify that alternative enclosing sequences eventuate in the same r value, which requires that we filter out some peculiar sequences that behave weirdly.

Turning to mechanics, similar finite evaluations can deposit mass and force densities upon local points q. The most unexpected of these local deposits is the notion of stress, whose oddities I'll consider in a moment (it represents the slippery customer most centrally responsible for the conceptual difficulties of classical mechanics). This is the general approach that is almost universally followed in a modern course in continuum mechanics. As far as the project of successfully capturing preexistent classical practice within a single axiomatized framework goes, the Truesdell-Noll system (or some of its close cousins) represents the best contender by far. At the end of this appendix, I'll return to the issue of its adequacy in this role.

The conceptually trickiest deposits of this localized nature are the notions of stress which are locally induced upon local points q by the following procedure. Following a diagnostic operation called an "Eulerian cut," we can carve out finite sub-volumes $\mathcal{R}_1$ of any size we'd like within an initial body $\mathcal{R}$. Arrayed around the boundary of $\mathcal{R}_1$ we will then encounter a distribution of so-called traction vectors that capture the contact forces acting across the interface between $\mathcal{R}_1$ and its $\mathcal{R}-\mathcal{R}_1$ complement. If we assume that we can continue making these Eulerian cuts within a nested sequence $\mathcal{R}_1$, $\mathcal{R}_2$, $\mathcal{R}_3$, ... that approach our central point q as closely as we'd like, we can take

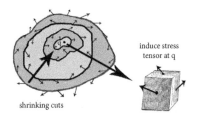

induce stress tensor at q

shrinking cuts

a limit and induce a density-like "stress" deposit on q that intuitively resembles a point that is pushed or pulled differently upon its various "sides" (a notion that is truly hard to picture!).

The simplest form of a stress of this kind is the notion of "internal pressure" which had been commonly invoked since antiquity without ever being coherently explicated. Indeed, Euler[21] himself would have rejected "his principle" as incoherent on the grounds that an "internal pressure" must ultimately stem from the outer surface forces that prevent one microscopic grain from penetrating another when they pressed into immediate contact (Euler 1768,

---

[21] Reint de Boer aptly comments:

> In his [*Mechanica*, Euler] also addressed the cut principle. However, his reflections have little in common with the formulations used today, which state that all balance equations are valid not only for the total body but also for every part which is imaginarily separated from the bulk body, where the freed interactions on the cut surfaces are to be attached as external interactions. In the course of his long research career, Euler never gave a clear statement of the cut principle. However, in his considerable works he employed the cut principle with mastery.
>
> (de Boer 2000, 7)

As de Boer also observes, Truesdell (1968) somewhat exaggerates Euler's actual contributions, which are more aptly attributed to Cauchy. But even here the historical record is murky because Cauchy employed different continuum physics methodologies within different applicational circumstances.

263–5).[22] Such expectations were responsible for long historical debate as to whether "hard atoms" must be postulated to prevent these $\mathcal{C}_1$, $\mathcal{C}_2$, $\mathcal{C}_3$,...dissections from collapsing into meaninglessness. But the modern vein of thinking crucial to modern continuum mechanics developments advises, "No, stay true to your Eulerian cut expectations and shrink your volumes all the way to a *stress* or *pressure*. It's merely that the end result will be a conceptual animal of an unexpected character (as a probability density is not truly a probability, a pressure is not truly a contact force). It took a very long time before this moral was fully absorbed by the physics community. Hertz himself employed the notion of stress quite deftly in his well-known theory of hardness (Hertz 1881) but would have brushed away any questions as to what he was talking about with an airy appeal to "limits."

I especially emphasize these modern considerations because the internal deficiencies of standard classical modeling practice that Hertz cogently identifies turn out to critically revolve around the notion of internal pressure, as we'll discover in Chapter 3. 4

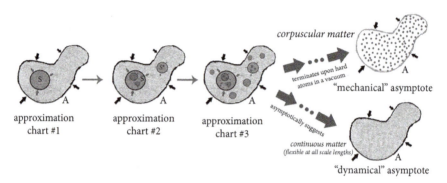

corpuscular matter

terminates upon hard atoms in a vacuum

"mechanical" asymptote

approximation chart #1

approximation chart #2

approximation chart #3

asymptotically suggests

continuous matter (flexible at all scale lengths)

"dynamical" asymptote

While on these topics, let me allow myself the luxury of a brief aside on the forward-looking attitudes that Kant strikes with respect to these basic issues of downward-heading descent in Kant (1786) (insofar as I have able to decipher his cryptic prose). He contrasts two basic approaches to the treatment of matter. The first of these he labels as "mechanical" and requires that any apparently continuous material must ultimately break into hard atoms in Euler's manner. But the second "dynamical" approach maintains that its true condition can only be asymptotically approached via an approximating sequence of finite bodies that influence one another through an admixture of what are now called "body" and "contact" forces. In the final limit, the provisional rigidities appearing in all of the approximating constructions vanish from view, leaving behind a smooth continuum innocent of any internal rigidities. I believe the contrast he has in mind can be aptly pictured as suggested: similar sequences of approximation chart that asymptotically support distinct limiting outcomes. If this represents a correct reading of his intentions, the "dynamical" situation comprises an astonishing anticipation of the modern thinking in continuum mechanics as it eventually developed in the wake of Hilbert's Sixth Problem.

---

[22] Euler also recognizes that such forces involve the contacts of many bodies simultaneously, and that simple two-particle principles such as Newton's coefficient of restitution rule cannot determine their respective magnitudes. He appeals to Maupertuis's Principle of Least Action as a means of resolving this difficulty. In doing so, Euler invokes a global minimization principle akin to those that Leibniz associates with "final causation" reasoning.

Kant plainly prefers the dynamical approach:

> Everything that relieves us of the need to resort to empty spaces is a real gain for natural science, for they give the imagination far too much freedom to make up by fabrication for the lack of any inner knowledge of nature. In the doctrine of nature, the absolutely empty and the absolutely dense are approximately what blind accident and blind fate are in metaphysical science, namely, an obstacle to the governance of reason, whereby it is either supplanted by fabrication or lulled to rest on the pillow of occult qualities.   (Kant 1786, 71)

But he further avers that this choice between the "mechanical" and the "dynamical" lies outside the reach of a priori metaphysical conclusion per se, because it concerns an infinitary condition that "can perfectly well be thought through reason, even though it cannot be made intuitive and constructed" (Kant 1786, 218). These barriers arise because it is only within the sequence of approximating "constructions" (viz., diagrams) that the Euclidean reasoning principles sanctioned as synthetic a priori within the pages of the *Critique of Pure Reason* can be properly applied.

Hertz plainly did not envision the direct path to continua just sketched and like most of his contemporaries presumed that we must sneak up on their expected behaviors by applying some type of limiting process to systems that only display a finite number of internal degrees of freedom. Virtually everyone in the physics community of the time would have agreed with this basic supposition, but most proposed approaching the task by crushing together discrete point mass "atoms," as indicated at the top of our diagram, leading to a continuous string in the alleged limit. However, Hertz instead suggests a limiting process in which short rigid bars pinned together with hinges or screws are progressively shortened until they lead to the same continuous string as a limit. Why does he prefer this policy? It is because the point mass "string" needs to be held together by action-at-a-distance attractions

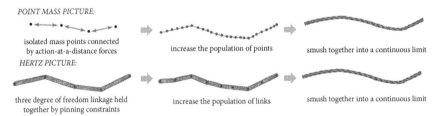

POINT MASS PICTURE:

isolated mass points connected by action-at-a-distance forces

increase the population of points

smush together into a continuous limit

HERTZ PICTURE:

three degree of freedom linkage held together by pinning constraints

increase the population of links

smush together into a continuous limit

whereas the short bars contraption is instead held together by pins, which represents a simple specimen of a so-called *geometrical constraint*. As I explain more fully in Chapter 4, Hertz believes that the central tensions encountered within standard modeling practice trace to the fact that the Newtonian-style forces are not conceptually compatible with the traction attachments operative within a constraint relationship (he is right about this, by the way). But his basic plan is to cure classical mechanics of its internal inconsistencies by doctrinal subtraction—remove conflicting claims until one reaches a self-consistent core suitable for axiomatization. So, the analyst faces a stark choice: Newtonian or contact forces? Hertz explicitly allows that some program of favoring the Newtonian picture might work out, but he doesn't see how it might work out in detail. Indeed, the point mass procedure will presumably start with the Euler's recipe as an initial gambit but would need to complete the picture by concretely specifying the "special force laws" needed to hold a string of point masses in a firm string-like array. But no plausible principles have ever been articulated that can accomplish that task in a satisfactory manner (Wilson 2009).

Hertz's well-motivated modeling choice then leads him into a thicket of unexpected expedients which will likely strike the reader as very peculiar. So peculiar, in fact, that the *Principles of Mechanics* is commonly misread precisely because many of Hertz's modern-day appreciators can't believe that he could possibly recommend doctrines as crazy as this. But Hertz is doggedly consistent in following out his announced program to the bitter end. As previously stated, none of these resultant oddities are especially pertinent to the central morals of this book, except to warn that the inflexible standards of righteous rigor can lead one inexorably into quite unattractive swamps.

In the remainder of this appendix, I'll sketch how some of these resulting oddities play out. To do so, let us borrow the convenient concept of a "mobility manifold" from modern robotics and the kinematics of mechanical movement. By this I merely the collection of all possible positions $\mathfrak{S}$ that a mechanism can potentially reach without breaking into pieces. I've drawn three of these possible positions for a so-called RPRPR device,[23] as well as marking by a dotted line the full mobility range of a specific cursor position on the device. As can be easily noticed, the geometry of these mobility spaces is apt to be quite complicated given its internal limits upon possible motions (note that the cursor's curve must terminate when the device's internal motions become blocked by its mechanical interconnections).

For Hertz, mechanical multiplicities of this interconnected character provide the prime data to which physical theory is responsible, for we can test any hypothesis as to its construction by simply twiddling with the device to find out the possible ways in which it can move (think of experimenting with a Rubik's cube, for example). He explains the conceptual centrality of the integrated mechanism as follows:

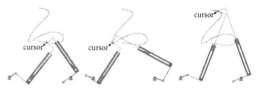

> Here the mechanics of a material system no longer appears as the expansion and complication of the mechanics of a single point; the latter, indeed, does not need independent investigation, or it only appears occasionally as a simplification and a special case. If it is urged that this simplicity is only artificial, we reply that in no other way can simple relations be secured than by artificial and well-considered adaptation of our ideas to the relations which have to be represented. But in this objection there may be involved the imputation that the mode of expression is not only artificial, but far-fetched and unnatural. To this we reply that there may be some justification for regarding the consideration of whole systems as being more natural and obvious than the consideration of single points. For, in reality, the material particle is simply an abstraction, whereas the material system is presented directly to us. All actual experience is obtained directly from systems; and it is only by processes of reasoning that we deduce conclusions as to possible experiences with single points. (Hertz 1894, 31)

Unfortunately, the terminology he employs ("system of material points") has sometimes misled commentators, because he is *not* referring to point masses in our Euler's recipe sense, but merely to labeling a particular rigid part $\mathfrak{R}$ by selecting convenient coordinates

---

[23] So called because of the types of linkages the mechanism utilizes: revolute-prismatic-revolute-prismatic-revolute.

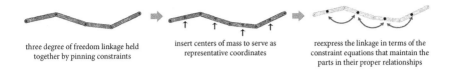

three degree of freedom linkage held together by pinning constraints

insert centers of mass to serve as representative coordinates

reexpress the linkage in terms of the constraint equations that maintain the parts in their proper relationships

for $\mathfrak{R}$'s location by appeal to center of mass and axis of inertia (if $\mathfrak{R}$ is allowed to rotate). Once the locations of these representative points have been established, we can easily fill out the locations of $\mathfrak{R}$'s remaining sectors from its associated constraint equations.[24]

As I understand Hertz's point of view, we develop our mobility space $\mathfrak{S}$ attributions to nature by "synthesizing the manifold of experience" using Euclidean geometry in Kant's approved manner. However—and this point will prove crucial to understanding Hertz's methodology—he proceeds to embed these complicated "mobility spaces" $\mathfrak{S}$ within larger varieties of covering manifold $\mathfrak{M}$ with respect to which $\mathfrak{S}$'s behaviors can be projected upon $\mathfrak{M}$ in a manner that is easier to articulate. This descriptive tactic had established itself as a valuable technique within the so-called "projective revolution" of nineteenth-century geometry long before Hertz appeared on the scene (it remains so to this day).[25] In this "revolution," it was recognized that the articulation of general geometry truths can be greatly facilitated by enlarging old-fashioned Euclidean geometry into a richer arena that contains supplementary "extension elements" such as points at infinity and points that reside at complex coordinates. Unlike the familiar non-Euclidean geometries, this projectivized enlargement does not present a rival to Euclidean orthodoxy but merely a helpful supplementation. It is in this spirit that Hertz proceeds to inflate his mobility spaces $\mathfrak{S}$ to richer manifolds $\mathfrak{M}$ over which he can articulate his single "dynamical law" quite simply (it is the only empirical law he needs). As it happens, this inflated $\mathfrak{M}$ possesses a natural geometry of its own that happens to be a non-Euclidean manifold of high dimension when considered in its own right. Hertz prefers saying that "it displays a geometrical analogy" for he insists that the only "true" geometry involved comprises the traditional Euclidean doctrines that deck out the space $\mathfrak{S}$ with its proper mobilities. Accordingly, an interconnected mechanical system's "mobilities" represent the prime data to which a suitable physics is responsible, but we can supplement these foundations with extra descriptive ingredients as long as we don't alter the verifiable mobility space $\mathfrak{S}$ that prompts the enlargement. The supplementary devices that Hertz adds in this allegedly innocuous manner convey him into very peculiar extended manifolds $\mathfrak{M}$. The fact that his latter-day readers have sometimes failed to recognize the oddity of his version of mechanics comes from failing to recognize the unusual methodological liberties he permits himself.

---

[24] J. G. Papastavridis writes:

> From such a continuum viewpoint, a particle is viewed not as the building block of matter, but as a rigid and rotationless body! As Hamel summarizes: "What one understands, in practice, by particle mechanics is none other than the theorem of the center of mass."
> (Papastavridis 2002, 100–1)

[25] The only mention of which I'm aware that Hertz offers on these tactical scores is a passing indication of an interest in "the meaning of imaginary quantities, of the infinitely small and infinitely large" in a letter of 1878 quoted in Nordmann (1998, 159). But these early investigations with respect to simplifying "extension elements" affected scientific thinking tremendously in the late nineteenth century and encouraged the notion that human thinking is distinguished from that of other animals precisely through its greater liberties with respect to creative thought (Wilson 2020). Hertz's willingness to ascribe implausible sub-infinitesimal hidden masses to external reality may result from an overly zealous cultivation of these "liberties."

Let me briskly outline some of the extension ingredient tactics he employs.

(1) Suppose we are faced with a simple member linkage whose connected bars weigh 2 kg, 3 kg and 0.5 kg respectively. He divides these parts into smaller pieces so that he can assign all of the smaller sectors a common unit mass of 1 following an appropriate choice of measure. As long as we introduce suitable constraint equations to hold the newly segmented pieces in rigid alignment with one another, this artificial inflation in S's formal degrees of freedom shouldn't alter its underlying mobilities one whit. But why on earth should Hertz ʌ̶ engage in this strange dissection?

(2) There are at least two reasons, one of which I'll discuss shortly, but the first is closely connected with the problems of measure that we discussed earlier in relation to continuum mechanics. We noted that regarded at the point level q, such a q could not coherently be assigned a positive mass, but only a mass *density*. However, if our target systems are compressible (as Hertz's ether must be), these densities will change from moment to moment. If so, how can we repress the conservation

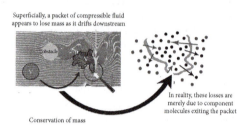

Superficially, a packet of compressible fluid appears to lose mass as it drifts downstream

obstacle

In reality, these losses are merely due to component molecules exiting the packet

Conservation of mass

of mass within such a setting? In a standard point mass universe, implementing the conservation requirement is straightforward: each particle retains its originally allotted mass wherever it strays. Mass will be conserved between two volumes V and V' if they both contain exactly the same mass points. But such bottom-up treatment[26] fails if an individual particle q within a flowing stream can alter its density as it moves ahead. Hertz's astonishing resolution of this difficulty is to cram infinitely many of his unit masses within a sub-infinitesimal region inside the "point" q.

(3) Hertz's most radical supplementation inserts a subterranean layer of additional material parts at a smaller-scale level (again, often sub-infinitesimal) to render the energetic behavior of his systems internally consistent. To do this, he resuscitates an old hypothesis to the effect that all energetic storage is actually kinetic in origin, albeit stored in movements that occur upon a microscopic level. Suppose we have two billiard balls that appear to attract and repeal one another through some potential energy function V associated with a conventional action-at-a-distance force. Relying upon some striking theorems on

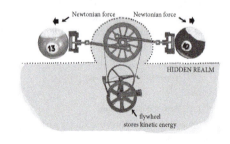

Newtonian force   Newtonian force

HIDDEN REALM

flywheel stores kinetic energy

"cyclic coordinates" developed by Helmholtz, J. J. Thomson, and Routh, Hertz argues that V's apparent action-at-a-distance activities can be perfectly mimicked by mechanically attaching the balls to an elaborate set of hidden flywheels whose kinetic energy spinnings supply the attractive and repulsive "juice" that feeds V's apparent repulsive and attractive

---

[26] In a modern continuum framework, mass conservation is installed in a top-down fashion with respect to every finite volume whose movements over time can be directly traced. At the infinitesimal q level, these arrangements induce a so-called "transport equation" for the mass density.

activities.[27] Of course, we lack any direct evidence for the presence of these hidden mechanisms, but Hertz observes that the molecular-based attractions that a force-based approach must postulate in order to hold a bead upon a wire are equally hypothetical.[28]

(4) Conventional potential-based mechanics dictates the manner in which a system's energetic expression shifts between kinetic and potential modes as it advances forward in time. Consider, for example, how a swinging pendulum continually shuttles energy between a purely potential storage (at the ends of its swings within a gravitational field) and kinetic movement (as the bob whizzes sweeps past its lowest point). The exact nature of this exchange is determined by the constraints that define the device (a bob that maintains a constant distance from a fulcrum) and some principle of extremal exchange. Hertz borrows his prototype from Gauss who relies upon a form of least squares calculation, which Arnold Sommerfeld explains as follows:

> [Gauss'] brief paper of 1829 entitled "On a New General Fundamental Principle of Mechanics," is concluded with the characteristic sentence, "It is quite remarkable that Nature modifies free motions incompatible with the necessary constraints in the same way in which the calculating mathematician uses least squares to bring into agreement results which are based on quantities connected to each other by necessary relations.". . .
>
> Clearly [his techniques] extend without change to non-holonomic constraints. Thus we are indeed confronted by a "new general fundamental principle of mechanics," as claimed by Gauss in the title of his paper. This fundamental principle is fully equivalent to d'Alembert's principle. Like the latter it is a *differential principle* in that it deals only with the present behavior of the system, not its future or past behavior. Here we do not need the rules of the calculus of variations, but only those of the ordinary differential calculus in the determination of the maxima and minima.
>
> (Sommerfeld 1943, 210–12)

By appealing to his supplementary hidden masses, Hertz converts Gauss's principle into a rule that governs how kinetic energy expression will shuttle about within a connected mechanical system, initially manifested within the bob's overt kinetic movements as it flies before a macroscopic observer and later within the faster rotations of its unseen subterranean flywheels.[29] Ernst Mach correctly observes that Hertz's general picture very much resembles that which Descartes embraced.

His further trick of dividing his mechanical parts into sectors of unit mass allows him to frame a "metric" based upon kinetic energy that allows him to recast the physics operative within $\mathfrak{M}$ in a so-called "phase space" portrait in which a single point representing a complete mechanical system moves about $\mathfrak{M}$ in a non-Euclidean manner whose local

---

[27] Hertz based these conclusions solely upon mathematical considerations; he did not attempt to implement these relationships in concrete terms. He was much criticized for this lapse in the ensuing commentary.

[28] Furthermore, we lack the special force laws required for this purpose, whereas Hertz is able to completely specify the only fundamental law he needs to move his systems forward in time. In his setting, all of the remaining tasks for which a Newtonian requires the specialized forces are addressed by crediting the affected system with an appropriate type of rigid body assembly. The fact that he never supplied such an assembly for classical gravitation (or the elastic ether, for that matter) bothered his critics, but he could at least claim that he manifestly lacked required ingredients in the manner of the Eulerian recipe.

[29] The basic idea is as the kinetic energy of the bob moves it forward, its linkages to the hidden flywheels will make them move as well, draining energy from the bob itself. Hertz "dynamical law" dictates that their rate of exchange must be extremal in the manner of the classical principle of virtual work.

curvatures enforce the desired requirements on energetic allocation. In this manner Hertz produces the enlarged manifold 𝔐 in which the single "fundamental law" he requires to govern the evolving behaviors found within the original mobility space 𝔖 can be more conveniently articulated.[30] Hertz underscores that fact that such an 𝔐 is constructed as a natural extension of the Kantian constructions encountered within the original 𝔖, rather than focusing upon 𝔐's non-Euclidean novelties in their own right.

In adopting these attitudes, Hertz conforms to a point of view that was common amongst the projective geometers in his period. Jakob Steiner, for example, believed that the novel projective extensions fill out the Platonic realm in which geometrical objects actually live:

> The present work contains the final results of a prolonged search for fundamental spatial properties which contain the germ of all theorems, porisms and problems of geometry so freely bequeathed to us by the past and the present. It must be possible to find for this host of disconnected properties a leading thread and common root that would give us a comprehensive and clear overview of the theorems and better insight into their distinguishing features and mutual relationships.... By a proper appropriation of a few fundamental relations one becomes master of the whole subject; order takes the place of chaos, one beholds how all parts fit naturally into each other, and arrange themselves serially in the most beautiful order, and how related parts combine into well-defined groups. In this manner one arrives, as it were, at the elements, which nature herself employs in order to endow figures with numberless properties with the utmost economy and simplicity.   (Steiner 1832, 315)

Hertz likewise hopes that the success of the reasoning practices captured within his 𝔐-manifold constructions also reveal an unvarnished world of systems-in-themselves lying behind the veil of our Kantian constructive tactics:

---

[30] Hertz adds an additional motivational comment that helps explain the exact manner in which his physics stops being synthetic a priori:

> As a second merit, although not a very important one, we specify the advantage of the form in which our mathematical mode of expression enables us to state the fundamental law. Without this we should have to split it up into Newton's first law and Gauss's principle of least constraint.... We cannot assert that nature always keeps a certain quantity, which we call constraint, as small as possible, without suggesting that this quantity signifies something which is for nature itself a constraint,—an uncomfortable feeling. We cannot assert that nature acts like a judicious calculator reducing his observation, without suggesting that deliberate intention underlies the action.... Our own fundamental law entirely avoids any such suggestions.... [I]t simply states a bare fact without any pretense of establishing it.
> (Hertz 1894, 30–2)

By his own admission, Wittgenstein was greatly influenced by *The Principles of Mechanics*, although it is hard to ascertain exactly how. I suggest that it may partially lie in the 𝔖 to 𝔐 relationship discussed here, where the analogous 𝔖 in the *Tractatus* corresponds to the actual multiplicities of real-life language (the set of variations we regard as meaningful) and 𝔐 comprises a non-distorting extension in which the 𝔖-level behaviors can be more conveniently captured. Wittgenstein's notoriously elusive "elementary propositions" may correspond to the independent coordinates that Hertz imposes upon the 𝔐 manifold, despite the fact that some of resulting paths will not qualify as "permissible trajectories" (because of non-holonomic constraint restrictions). More generally, Wittgenstein appears to have taken to heart Hertz's advice that behaviors should be analyzed as connected systems rather than by attending to individual words or sentences apart from the pragmatic employment.

Josh Eisenthal has been investigating some of these potential parallelisms in a similar vein, although I am unsure how many of my suggested glosses he would accept.

If we try to understand the motions of bodies around us, and to refer them to simple and clear rules, paying attention only to what can be directly observed, our attempt will in general fail. We soon become aware that the totality of things visible and tangible do not form a universe conformable to law, in which the same results always follow from the same conditions. We become convinced that the manifold of the actual universe must be greater than the manifold of the universe which is directly revealed to us by our senses. If we wish to obtain an image of the universe which shall be well-rounded, complete, and conformable to law, we have to presuppose, behind the things which we see, other, invisible things—to imagine confederates concealed beyond the limits of our senses.... We are free to assume that this hidden something is nought else than motion and mass again,—motion and mass which differ from the visible ones not in themselves but in relation to us and to our usual means of perception. Now this mode of conception is just our hypothesis. We assume that it is possible to conjoin with the visible masses of the universe other masses obeying the same laws, and of such a kind that the whole thereby becomes intelligible and conformable to law. We assume this to be possible everywhere and in all cases, and that there are no causes whatever of the phenomena other than those hereby admitted. This is the leading thought from which we start.   (Hertz 1894, 25–6)

However, he is wary of claiming too much, partially because an alternative development might yield a different physical picture and partially because some of his postulated ingredients presumably lie permanently beyond the reach of empirical inquiry (e.g., at sub-infinitesimal levels). I'll return to some of the issues attendant upon such forms of "scientific realism" in Chapter 6.

Before concluding this summary of Hertz's concrete procedures, let me briefly indicate his objections to the second of the "three images" of classical mechanics to which he often alludes but which I have skipped over. In this second "image" Hertz has in mind a mechanics squarely founded upon the conservation of energy as encapsulated within Hamilton's variational principle. Hertz's objection to this scheme is that it doesn't deal with so-called non-holonomic constraints (such as rolling or ice skating upon a plane) in the right way, a fact that was already known to Lagrange himself who favored a Gauss-like virtual work approach to mechanics over his own Laplacian function treatment (which confronts allied problems with respect to these constraints). Hertz further objected to the manner in which the temporally extended variations utilized in these energetic approaches mystically suggest calculation upon nature's part.

In retrospect, the strangest aspect of the philosophical enthusiasm for axiomatic purification that Hertz's book inspired lies in the fact its latter-day enthusiasts have commonly presumed that the "image" that Hertz regarded as the least promising (a mechanics centered upon the Newtonian concept of force) can be rendered fully coherent after a minimal degree of conceptual adjustment. This presumption is plausible only if one is willing to presume that the bare skeleton of the Eulerian recipe sketched above comprises a "complete theory," ignoring the fact that real-life applications (like beads on a wire) utilize constraints in lieu of the missing "special force laws." Failing to recognize this crucial deficiency (which would have been impossible if the parties in question had read Hertz's introduction carefully), it becomes easy to axiomatize the rest of the Eulerian recipe in the superficial manner in McKinsey, Sugar, and Suppes (1953).[31] Because this mathematical framework is easy to understand (insofar as it reaches), this

---

[31] This treatment is severely, but justly, criticized in Truesdell (1984).

perversion of Hertz's original project has encouraged the presumption that "the general format in which theories operate is well understood, however intricate its internal details might be." That, of course, is the methodological tenet upon which most of the "ersatz rigor" presuppositions of contemporary philosophical method rely. None of these unfortunate "theory T" distortions should be laid at the feet of Hertz himself, but arise only later, at the hands of Rudolf Carnap and the other "logicizing" methodologists of his era. All of them appear to have completely forgotten about the delicate considerations that Hertz more squarely confronted.

As I've already stressed, all of these arcane details are extraneous to the central project of this book, which simply documents how the "ersatz rigor" of single-level axiomatic organization came to dominate diagnostic effort within contemporary philosophy. However, a firmer awareness of the developmental details sketched here might assist that specific vein of modern "metaphysical" inquiry that has resuscitated some of classical mechanics' venerable classical dilemmas with respect to matter and force and has unwisely presumed that they can be easily vanquished with a bit of armchair cogitation. They are better advised to either drink more deeply of Hertz or taste *not* that *Pierian* spring.[32]

In these respects, Hertz's *Principles of Mechanics* provides a wonderful illustration of how the purest veins of refined thinking, logically pursued, can lead one into dreadful impasses, for there is no doubt that the framework he embraces would bring effective modeling practice to a grinding halt, its gears and wheels coated in the unworkable grime of needing to replace Newtonian forces by constraints. But we have also observed that subsequently vanquished mathematical obstacles blocked his path to a far more appealing approach in which plausible axioms for continuous materials are directly specifying by employing the top-down policies of modern measure theory. As indicated above, Hilbert sparked interest in such an approach in his 6th Problem which subsequently led to a long series of studies initiated by Hilbert's student Georg Hamel and continued into the 1950s by Clifford Truesdell and Walter Noll.[33] This is the canonical framework in which "classical physics" is now taught to modern engineers. Indeed, Truesdell regarded his team's efforts as aiming towards exactly the flavor of closed-world axiomatics for thermomechanics that Hertz originally envisioned. His emphasis on internally enclosed rigor often resembles the haughty disdain in which Pierre Duhem dismisses the "shallow thinking" of British authors such as Kelvin and Tait, who are less dogmatic in their presumptions with respect to "foundations."

More recent experts such as Gérald Maugin generally reject Truesdell's axiomatic presumptions as unduly "purist":

> The bias, shortcomings and audacity [of Truesdell's approach] illustrates the elegance and temptations of all formal logico-deductive approaches: it is very attractive.
>
> (Maugin 1998, 12)[34]

---

[32] Pope properly continues: "More advanced, behold with strange surprise/ New distant scenes of endless science rise!"

[33] Perhaps the best single source for this material is Truesdell (1977), although most of its contents now permeate most modern books on the topic. It should be noted that these efforts do not embrace situations in which boundary and interfacial regions are altered by interior stresses in the manners we shall discuss later. Interesting attempts to bring these processes into the general ambit of Noll-style mechanics can be found in Gurtin (1999) or Maugin (2011).

[34] See Rivlin (1997). Of this divergence, Michel Destrade, Jeremiah Murphy, and Giuseppe Saccomandi write:

The central reason for this shift in attitude is closely tied to the need for non-standard forms of continua such as the Cosserat media we discussed in connection with ether models or laminates that contain adjusting boundaries. In the newer work, appropriate behavioral postulates are often reached by "homogenizing" some lower-scale modeling in the general vein we shall discuss in Chapter 4. Indeed, some of these supportive submodels utilize point particle constructions for their "constitutive equation" guidance, again illustrating the Escher staircase-like dependencies that supply classical mechanical practice with its wide-ranging modeling capacities but which also produce puzzling contextual clashes as a side effect of that very success. And why not? The only reason we might have for anticipating complete foundational purity within a classical mechanical frame is if we could be assured that nature itself manifests firm doctrinal preferences along those lines. But no; when we attend closely to nature's lower-scale minutiae, it begins to mumble about quantum mechanics instead.

After the Second World War, a group of talented mechanicians and mathematicians worked hard at transforming continuum mechanics into a Rational Mechanics program. Ronald Rivlin and Clifford Truesdell were undoubtedly the leaders of this program. In the early days, Rivlin and Truesdell had cordial relations based on mutual respect, but later on their relationship deteriorated, as some of Truesdell's followers were encouraged to develop mechanics as a pure axiomatic subject in the spirit of David Hilbert. This was a departure from the spirit of rational mechanics as pursued by Paul Appel in France or Tullio Levi-Civita in Italy, where the challenge was to combine deep theoretical analysis with concrete practicality. If only this clash of characters and schools between Rivlin and Truesdell had been avoided, then continuum mechanics would have enjoyed an unassailable advantage in researchers and resources.

(Destrade, Murphy, and Saccomandi 2019, 3)

It is Rivlin's point of view that largely prevails today.

4

The Mystery of Physics 101

> But the answer we want is not really an answer to this question. It is
> not by finding out more and fresh relations and connections that it
> can be answered; but by removing the contradictions existing between
> those already known, and perhaps by reducing their number. When
> these painful contradictions are removed, the question as to the
> nature of force will not have been answered; but our minds, no longer
> vexed, will cease to ask illegitimate questions.
>
> Heinrich Hertz (Hertz 1894, 7–8)

(i)

In the previous chapter, we noted Hertz's complaint that "it is exceedingly difficult
to expound to thoughtful hearers the very introduction to mechanics without
being occasionally embarrassed. The renowned hydrodynamicist Horace Lamb[1]
confesses to a similar discomfort:

> To a student still at the threshold of the subject the most important thing is that
> he should acquire as rapidly as possible a system of rules that he can apply
> without hesitation, and, so far as his mathematical powers will allow, with
> success, to any dynamical question in
> which he will be interested. From this
> point of view it is legitimate, in expound-
> ing the subject, to take advantage of
> whatever prepossessions he may have as
> are serviceable, whilst warning him
> against others which may be misleading.
> This is the course which [is] attempted...
> in most elementary accounts of the sub-
> ject. But if at a later stage the student,
> casting his glance backwards, proceeds
> to analyze more closely the fundamental

Lamb

[1] Lamb supervised Wittgenstein's aeronautical studies at Manchester.

Imitation of Rigor: An Alternative History of Analytic Philosophy. Mark Wilson, Oxford University Press. © Mark Wilson 2022.
DOI: 10.1093/oso/9780192896469.003.0004

principles as they have been delivered to him, he may become aware that there is something unsatisfactory about them from a formal, and even from a logical, standpoint.... If the student's intellectual history follows the normal course he may probably, after a few unsuccessful struggles, come to the conclusion that the principles which he is virtually, though not altogether expressly, employing must be essentially sound, since they invariably lead to correct results, but that they have somehow not found precise and consistent formulation in the textbooks. If he chooses to rest content in this persuasion, deeming that form and presentation are, after all, secondary matters, he may perhaps find satisfaction in the reflection that he is in much the same case as the great masters of the subject: Newton, Euler, d'Alembert, Lagrange, Maxwell, Thompson and Tait (to name only a few), whose expositions, whenever they do not glide hastily over preliminaries, are all open, more or less, to the kind of criticism which has been referred to. The student, however, whose interest does not lie solely in the applications, may naturally ask whether some less assailable theoretical basis cannot be provided for a science which claims to be exact, and has a long record of verified deductions to its credit.   (Lamb 1923, 345–6)

Unfortunately, Lamb's suggestions for rectifying this situation are not adequate either, for reasons like those we shall rehearse here.

I call these puzzles "The Mystery of Physics 101" because students within an opening-level university course frequently become mystified by the subject's abrupt twists and turns. I was certainly one of these. In high school, courtesy of an older brother who was training to become a philosopher, I had absorbed various 1950s era primers composed in what was then called a "logical empiricist" mode. Yet it was clear that the toy "theories" in which these books trafficked could not comprise the stuff of which real science is made,[2] so I eagerly awaited a proper exposure to a real-life "theory T" of the sort they described. Indeed, matters appeared to start off well enough from an axiomatic perspective that first week with "Newton's Laws of Motion" (although that "action = reaction" business seemed oddly formulated). But my comprehension went swiftly downhill when we turned to beads sliding along wires and allied topics the second week in. I knew little actual science but was eminently

The Mystery of Physics 101

---

[2] An alleged "law of nature": "phosphorus smells like garlic." This disconcerting specimen is extracted from Scheffler (1963), which was one of the philosophical primers that my brother brought home from college.

"logical" thanks to my brother's books. And I could not see, for the life of me, how these conclusions about beads and wires followed in any clear sense from the "axioms" of Week 1. "Of course, they do," snarled my unsympathetic instructor, "concentrate upon your homework."

Such is the unsettling manner in which substantive conceptual problems originally emerge. Their minor surface disharmonies are symptomatic of significant tectonic tensions underneath. My Physics 101 confusions with respect to beads siding on a wire are secretly symptomatic of deeper conceptual disharmonies that have historically shaped metaphysics and the development of science in significant ways. These are the entanglements that we shall trace in this book. How can science prove successful if it utilizes its mathematical tools in this seemingly inconsistent manner? This is the sort of quizzical inquiry I had in mind when I wrote that "even metaphysicians start small" in our opening chapter. As it happens, Heinrich Hertz applied his admirable powers of intellectual discernment to this very problem in his *Principles*. The purpose of this chapter is to explain what he saw.

As we noted in the previous chapter, Hertz does not claim that the notion of "force" at the center of these mismatches is inherently "metaphysical" or muddled, but merely that established practice has loaded the term with an excessive number of descriptive demands. Hertz believes that a conceptual reformer should examine the inferential machinery embodied within a problematic descriptive practice and tinker with its interconnections until these parts no longer grind against one another inconsistently. Such emendations are inevitable in any progressively evolved descriptive practice. So it is little wonder that perfectly functioning machinery rarely emerges as their accumulated product.

Drawing upon Hertz, David Hilbert captures these developmental themes as follows:

The buildings of science are not erected the way a residential property is, where the retaining walls are put in place before one moves on to the construction and expansion of living quarters. Science prefers to get inhabitable spaces ready as quickly as possible to conduct its business. Only afterwards, when it turns out that the loosely and unevenly laid foundations cannot carry the weight of some additions to the living quarters, does science get around to support and secure those foundations. This is not a deficiency but rather the correct and healthy development.
(Hilbert 1905, 103–4)

Hilbert

This "holistic" conception of conceptual repair was widely shared among many of Hertz's scientist/philosopher contemporaries. Writing in this vein (and also acknowledging Hertz explicitly), Ernst Cassirer writes:

> Physics is not a machine that can be taken apart. One cannot test every piece individually and wait until it has stood this sort of testing before putting it into the system. Physical science is a system that must be accepted as a self-contained whole, an organism of which no single part can be made to work without all the others, even those farthest removed, becoming active—some more, some less, but all in some precise degree. When the functioning is blocked in any way the physicist must try to guess from the behavior of the whole which part is at fault and in need of repair, with no possibility of isolating and testing it separately. Given a watch that will not go, the watchmaker removes all its parts, examining them one by one until he finds which was out of place or broken, but the physician cannot dissect a patient in order to make a diagnosis; he must guess at the seat and cause of the disease merely by studying a disturbance that affects the whole body. The physicist resembles the latter when he has a defective theory before him which he has to improve.   (Cassirer 1950, 113–14)

Implicit within these developmental remarks is the psychological observation that such enlargements generally unfold without the affected parties evincing any cognitive awareness that the "meanings" of their terminologies have somehow shifted: the only signal that something unusual has occurred is revealed by the fact that on occasion A we find ourselves asserting S whereas on occasion B we instead claim not S. Commonly, it is not apparent that either of these clashing claims is mistaken; each will seem like "exactly the right thing to say" on the appropriate occasion of usage. Direct musing upon the internal conceptual character of the word "force" in isolation from the reasoning policies with which it is associated will rarely produce useful results. Pragmatic holists like Hertz and Cassirer accordingly reject the picture of "conceptual analysis" that we associated with Russell (1914) in the previous chapter; in itself, the individual word "force" does not carry a defective "meaning" that renders it internally problematic. We'll revisit these methodological disparities with respect to the tasks of "conceptual analysis" in Chapter 7.

(ii)

These are the conceptual circumstances that Hertz confronts within *The Principles of Mechanics*: historical developments have deposited an overloaded docket of doctrines on his doorstep, in which action-at-a-distance forces between point masses comingle freely with the larger-scale restrictions on movement which the "constraint forces" acting against the moving bead represent (I'll amplify upon this comment momentarily). As we've seen, reasoning admixtures of this type naturally arise as the organic products of free-wheeling scientific improvement. But their contents must eventually submit to a more disciplined form of doctrinal policing. At some point significant repairs will be needed, for so long as these subterranean conflicts are left unresolved, we leave ourselves vulnerable to potentially damaging tangles, for unmonitored contradictions can trick us into accepting anything.[3]

Hertz's remedy is to isolate a reduced and self-consistent form of axiomatic housing from which these conflicting strands have been carefully excised. As we have already previewed, Hertz rather unexpectedly favors the "constraint" side of the beads-versus-constraint ledger in carrying out this doctrinal purging. If so, what are the basic entities with which he must work in carrying this revisionary scheme forward? *Answer*: the assemblies that we can construct by binding together an ensemble of rigid parts in the manner of clockwork (viz., gears engaged with neighboring gears; balls that roll or slide along a groove in metal plate and the like). These various forms of relative movements are officially called "*constraints*" and can be conveniently registered in simple algebraic or differential relationship terms, which capture the intuitive notions that "rod A must rotate freely around rod B because of the pin placed at point p" and/or "wheel A is obliged to roll freely across the rigid plane B." The upshot is a world of mechanical gizmos whose interlaced parts roll or slide across their neighbors' surfaces in the manner of clockwork or a steam shovel. If we ignore frictional effects, we can determine the "kinematics" (= the possible movements)[4] of these connected assemblies through geometry alone, without any appeal to

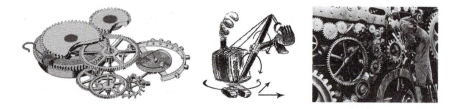

<hr/>

[3] Elementary logic tells us that any slight syntactic contradiction can be milked into a completely valid argument in favor of any conclusion we like. E.g., Assume P and not P. From P, infer the weaker P or Q. Using not P, cut away the "bad part" of P or Q, obtaining Q, where Q can be any statement whatever.

[4] Note, however, that we have not yet equipped these assemblies with a "dynamics" that will carry them forward in time. Hertz provides a single, non-Newtonian law to this purpose in his completed scheme, but I will postpone consideration of how this is accomplished until our supplementary Chapter 6.

Newtonian-style "forces."[5] As we noted in Chapter 2, these assemblies (which we'll usually label as "mechanisms" or "machines") comprise the natural inhabitants of the "clockwork universes" of traditional expectation, although they do not readily suit the contours that Newton actually outlines in his own *Principia* (Wilson 2020a).

Accordingly understood, the central ingredients within Hertz's reconstructed mechanics consist entirely of integrated gizmos such as gear trains and steam shovels, whose allowable movements form into collections (or "manifolds") that workers in robotics sometimes call "mobility spaces." The "degrees of freedom" (= ranges of possible wiggling) proper to these devices are determined by their geometrical interconnections, rather than comprising the unconstrained x, y, and z movements that are available to a swarm of unattached point masses. As such, the internal geometries of these "mobility spaces" are considerably more complicated mathematically than the comparable "phase spaces" familiar from point mass mechanics. Although he doesn't employ this terminology, these "mobility spaces" play a central role in Hertz's conception of mechanics. Since these details will not affect our central discussion much, I'll reserve to our optional appendix a sketch of the unexpected manners in which Hertz eliminates residual appeals to Newtonian action-at-a-distance forces from these mechanism-styled universes. Although this tactic is often loosely characterized as "doing without forces," it is more aptly described as "doing without Newtonian-style forces," for "constraint forces" that hold a bead to a wire remain fully in place (although Hertz rarely employs this standard terminology).

For want of a better term, I shall call this scheme a "machine-based form of rigid body physics." As Ernst Mach and others have accurately noted, it represents essentially the same conception of the mechanical universe as Descartes favored, although the corresponding mathematical details are considerably different (Mach 1883; Wilson 2021). This approach should be sharply distinguished from another traditional branch of mechanics (sometimes called "stereomechanics") that allows detached geometrical entities such as billiard ball objects to interact with one another via impactive contact. We shall discuss these behaviors in greater detail in Chapter 6 and will merely remark that impactive contacts of this type are sternly forbidden within Hertz's universe of smoothy interconnections (Hertz 1894, 255).

This conception notion of mechanical devices held together by geometrical constraints alone is so immediately appealing that it's a shame that later commentators have often failed to recognize that it comprises the target "physical ontology" that Hertz has intuitively in mind. As noted before, the physicists who

---

[5] In Hertz's setting, these constraint relationships completely obviate any necessity for "special force laws" in the manner of Euler's recipe. He requires only a general "dynamical law" (based upon Gauss's least work principle) to move his mechanisms forward in time autonomously.

commented upon *Principles* immediately following its publication correctly identified its intended objectives, but later writers have frequently misread Hertz in a point mass manner, for the amnesic reasons surveyed in Chapter 3. To be sure, his austere writing style hasn't helped matters in these regards, as I shall discuss further in the appendix below.

As just remarked, these "mobility spaces" delineate an identifiable collection of "possible worlds" that differ considerably in character from the comparable constructions that arise as "phase spaces" within an Euler's recipe approach based upon point masses. Eliding both groupings together as "classical possible worlds" only encourages vagueness and loose reasoning, although this represents an extremely common practice within contemporary philosophical discussion.[6] In the mouths of physicists, sometimes the phrase "classical" merely connotes "not utilizing quantal or relativistic ingredients," but appeals of this kind plainly do not isolate clearly articulated collections of "possible worlds." Considerable philosophical labor has been devoted to a flurry of "formal results" with respect to traditional topics such as "is classical mechanics deterministic or not?" that have not specified their presumptive bases adequately. In many cases, the significance of these alleged "theorems" remains entirely opaque due to the murky way in which their appeals to "classical model" have been rendered. Such premature projects in "conceptual analysis" should be eschewed until a sounder resolution of the basic ambiguities of which Hertz rightfully complains has been established.

This habit of "getting ahead of your skis" with unhelpful "theorems" is emblematic of the "ersatz rigor" of which this volume complains. The fact that folks have often wondered about the "determinism of classical physics" in a loose manner doesn't license the unfocused tilting at windmills that appear when would-be rigorists fail to adequately identify their intended objectives beforehand (Wilson 2009). In contrast, Hertz's close scrutiny of applicational detail illustrates how a sounder form of conceptual diagnosis should begin. In the remainder of this chapter, we will examine the manner in which Hertz carries out his detective work.

This endorsement of Hertz's diagnostic insights does not entail that we should accept the remedies that he proposes for alleviating these performance flaws. As we've seen, he finds his suggested correctives within a sternly regimented axiom scheme based upon a rigorous pruning of the tangled array of doctrines that prior practice has bequeathed us. Hertz readily allows that this purging of conflicting "fundamentals" might be achieved in a variety of ways that may embrace the very conceptual categories (i.e., "Newtonian force") that he himself eschews. This

---

[6] One frequently encounters prose like this (drawn from a contemporary author I shall leave unnamed): "Consider a classical universe in which various colored blocks interact with one another," with the tacit presumption that such structures appear as "models" (in the logician's sense) of some unsupplied set of "axioms for classical mechanics." Philosophical insight is not assisted by incoherent appeals to non-existent doctrines of this type.

diagnostic tolerance[7] directly stems from his "holist" conception of the tangled origins of conceptual conflict. Nothing internal to the individual terminologies of traditional mechanics reveals any inherent ideological flaw; their employments simply do not work together ably in collective enterprise.

However, does this natural process of exclusionary winnowing represent the proper way to tame the conceptual tensions of which Hertz rightfully complains? After all, his dogged pursuit of hygienically cleansed axiomatic "purity" paints him into a very peculiar (and completely unworkable) doctrinal corner (as the survey provided in Chapter 2's appendix vividly illustrates). Chapter 5 will explicate an entirely different manner of responding to his dilemmas.

We might observe that the mere assumption that conceptual confusions should inevitably submit to axiomatic correctives encapsulates a tacit form of "metaphysical" demand of exactly the kind that writers like Hertz and Mach correctly regarded as detrimental to scientific advance. In particular, the requirement of uniform axiomatization places a number of "small metaphysics" demands upon the capacities of mathematical technique that deserve closer philosophical scrutiny (some of which we will attempt in Chapter 7). I find it rather amusing that the fiercest modern critics of "pernicious metaphysics" (Carnap and his followers) secretly rely upon axiomatics-based presumptions that should themselves be regarded as "metaphysical" in an excessively speculative vein. "Back to the 'small metaphysics' details first!," I say. But I will delay these criticisms further until we finish our survey of "The Mystery of Physics 101."

(iii)

At this juncture, I confront an expository dilemma. Although the critique that Hertz offers with respect to established mechanical practice is exceptionally penetrating, his own prose is not ideally pellucid on this same score. Even his ablest commentators have often misidentified the foci of his complaints (which center upon some overlooked factors involving "pressure" and Newton's Third Law of Motion). Such nuanced subtleties are only to be expected, for conceptual conflicts quietly worm their ways into a descriptive practice largely by remaining minute and muddied by misapprehensions of a wider canvass. These are the reasons why Hertz's historical puzzlements with respect to "force" offer a wonderful illustration of how real life conceptual problems typically emerge in situ. But this monograph also hopes to critique prevailing standards of ersatz rigor for the benefit of a wider philosophical audience, not merely for the few worthies familiar with the applicational delicacies of Newton's Third Law of Motion. I will

---

[7] As we later see, Rudolf Carnap's celebrated "principle of tolerance" undoubtedly traces to these thematic antecedents (Carus 2007).

accordingly divide our discussion into two parts, the first of which will provide a "quick and dirty" summary of Hertz's complaints that should suffice for our object-ives in the remainder of the book. The finer details of Hertz's developed discussion are relegated to the appendix, largely because its penetrating insights have not been properly acknowledged within the standard commentaries. Most of my readers can skip this supplementary aspect of our discussion, unless they are eager to witness at close range the inconspicuous pathways along which the clever rodents of conceptual confusion irrepressibly invade our domestic reasoning policies.

Here's the "quick and dirty" moral: a subtle ambiguity within our standard appeals to "force" renders the usual expression of the Conservation of Energy incoherent. Hertz commences his analysis as follows:

> In [Newton's] laws of motion, force was a cause of motion, and was present before the motion.... "Force" is introduced as the cause of motion, existing before motion and independently of it.... Can we, without confusing our ideas, suddenly begin to speak of forces which arise through motion, which are a consequence of motion? Can we behave as if we had already asserted anything about forces of this new kind in our laws, as if by calling them forces we could invest them with the properties of forces? These questions must clearly be answered in the negative.   (Hertz 1894, 6)

To see what he has in mind, let us return to our bead on a wire, subject to a uniform gravitational force. Locally, we can approximate this force with a simple directed vector $\mathbf{g}$ derived from the classic Newtonian formulation enshrined within the Eulerian recipe of the previous chapter as:

$$m_i d^2 \mathbf{q}_i(t) dt^2 = \Sigma_j (G.m_i.m_j)/|\mathbf{q}_i(t) - \mathbf{q}_j(t)|^2.$$

As such, the gravitational force $\mathbf{f_g}$ specified on the righthand side serves as the primary "cause" of the acceleration or deceleration that the bead p displays as it travels along the wire W. But this doesn't comprise the full story, because the unchanging shape of the wire sup-plies an opposing constraint force $\mathbf{f_c}$ that exactly cancels any bead move-ment that would carry it inside the

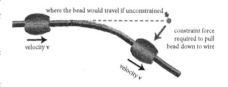

wire W. Operating in tandem, these two forces produce an effective resultant force $\mathbf{f_g}|\cos \varphi$ that varies in strength according to the angle that W makes with respect to the undiluted $\mathbf{g}$. The bead experiences the full strength of $\mathbf{g}$ only when it is moving vertically down the wire W. It will not respond to $\mathbf{g}$ at all if it happens to be confined to a perfectly horizontal portion of W (because the opposing con-straint force $\mathbf{f_c}$ completely nullifies $\mathbf{f_g}$'s influence). In the usual jargon of mechanics, the applied **g** force "performs no work" upon the bead in the latter circumstances. So here is the crucial distinction that Hertz draws in the quotation above: the f_g component directly derives from a Newtonian special force law that determines f_g's strength by a rule that reflects its condition in a manner "existing before motion and independently of it." But, this same characterization fails with respect to the opposing constraint f_c for its strength cannot be determined until we know how rapidly p travels across the pertinent stretch of wire. Why is that? f_c's strength must pull the bead p back to the exact position along W where it would have traveled had it maintained a constant velocity with respect to W. The faster W's velocity happens to be at this moment, the stronger the restoring force f_c must be. As such, the strength and magnitude of f_c arises as "a consequence of its motion" in the sense that the demand that the bead remain on the wire W serves as the causal determinant of what f_c's strength will be at any selected moment. Indeed, factors like f_c are commonly designated as "reaction forces" or "forces of constraint" precisely because their strengths must be computed in this after-the-fact manner rather in the before-the-fact manner of the Newtonian f_g. Hertz correctly notes that these "constraint forces" are nowhere mentioned in a Eulerian recipe scheme concerned with various varieties of "special forces." The "reaction force" f_c somehow sneaked in through the backdoor along with the wire W. I could not have analyzed the situation as crisply as Hertz, but this is the nub of the "illogical" concerns that troubled my juvenile axiomatic instincts back in the days of Physics 101.

(iv)

So that's the underlying conceptual tension, but why is it important? I'll address this practical question shortly. However, I'd first like to offer a variant diagnosis of our bead and wire circumstances that nicely presages the point of view we shall develop in Chapter 5. But insofar as our chief objectives go, this discussion is entirely supernumerary, and uninterested readers may advance directly to the

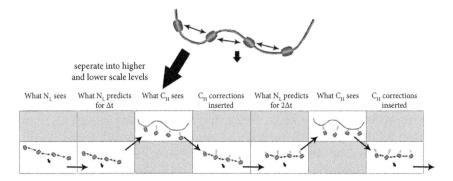

✓

section (iv). What I hope to provide is an account of how Hertz's force distinctions align with the so-called "Lagrangian multiplier" techniques one finds in textbooks that deal with imposed constraints successfully. I will attempt to explicate this methodology in non-technical terms by utilizing a form of military metaphor (if this narrative endeavor fails, no matter; just move forward to section (iv)). Let's first complicate our previous circumstances slightly by placing a greater array of beads upon our wire (I added four in my illustration) and subjecting them all to the original imposed gravitational field $\mathbf{g}$ and some additional action-at-a-distance force $\mathbf{f_e}$ such as electrostatic attraction. To capture Hertz's division between force types, let us further divide our problem into two scale levels, corresponding to the short characteristic length of the bead Δ_B and to the longer characteristic length of the wire Δ_W. I am invoking this intuitive distinction in scale lengths because I want to metaphorically align our two species of "force" with commanding officers of different ranks, a lower-scale Newtonian force sergeant S_B and a higher-scale constraint force general G_W. The lower-scale sergeant can see how the two Newtonian forces (marked in black) pull the beads together under her immediate command but can't observe how their placements relate to the longer wire W itself, which represents a larger and more complicated object than the little beads she watches over. So S_B leaves the task of monitoring the larger landscape of the wire to the attention of her commanding general G_W who keeps track of the obligation to keep the beads sliding along the wire W. This officer G_W can ably survey how the platoon of beads are arrayed with respect to the wire but can't directly observe the small-scale Newtonian (black) forces that connect them together. To determine how the beads will move forward in time, our two officers divide the task into two parts. Without worrying about the wire, the sergeant S_B begins the instructions by consulting her Euler's recipe rule book to decide how her squad of beads should march forward in a small unit of time Δt under the presumption that no wire W is present. These predicted movements are marked in the first two rectangles of our chart. But these movements will pull our beads off the wire, prompting the higher-scale general G_W monitoring the situation to bark a corrective order down to S_B, "Move your squad back onto that wire!" S_B will obediently include the general's constraint force arrows (white) into her revised orders. In so doing, she will successfully push the beads back onto the wire while otherwise conforming to her Euler's recipe rule book as closely as possible. But what should our sergeant S_B order with respect to a second advance in time $\Delta t + \Delta t$? If she consults her rule book and includes the constraint force instructions she received from the general a moment before, she will order her platoon into the fifth rectangle on our chart. But this is not satisfactory, because the beads will have moved off the wire once again. Seeing this, the supervising general G_W must correct the discrepancies by issuing fresh constraint orders with revised magnitudes and directions. And back and forth these cross-scalar accommodations will zig-zag, until the two commanders agree upon a compromised path forward for the beads. Note that Newtonian force sergeant S_B frames her original marching orders *before any motion* occurs by consulting her Euler's recipe rule book, whereas the constraint force general G_W issues his countervailing instructions by *scrutinizing the motions* ordered by the sergeant and correcting their lapses. Hertz's division between the "forces which arise through motion and those that are a consequence of motion" reflects this difference in character of the two kinds of order we witness within our military analogy. The conceptual tensions between these basic "force" classes stems from the fact that the magnitudes of the general's constraint forces need to be sensitive to the velocities of the beads, whereas the sergeant's Euler recipe prescriptions can only depend upon their relative displacements.[8] These peculiar distinctions in "causal" direction will resurface in a much wider context in Chapter 6.

Although I have explicated these computational distinctions in quasi-military terms, they correspond quite exactly to the differences in underlying mathematical character between the $F = ma$ differential equations that S_B employs and the (generally)[9] algebraic formulas that the constraint focused G_W consults. The back-and-forth consultations between our two commanders accurately reflect the so-called "Lagrange multiplier" tactics that practitioners employ to bring their $F = ma$ equations into alignment with the imposed constraints.[10] When constraints and Newtonian forces co-associate within a modeling effort, they will be governed by equations of distinctly different mathematical characters. But, to presage a moral that I will repeat later: not all equations act alike!

Calculations that operate in this back-and-forth character are said to proceed through *feedback corrections*, and we shall frequently encounter similar reasoning arrangements later in our narrative. Indeed, the secret agenda that motivates our present discussion is to presage the manner in which Hertz's observations with respect to incompatible force types will become recast as a component of "multiscalar architecture" in Chapter 5.

(v)

To return to an earlier question, so what's the big deal? Why should it matter that classical mechanics employs forces of these divergent types? My freshman instructor correctly presumed that if I couldn't complete a simple bead-on-a-wire

---

[8] As we shall shortly see, these tighter demands upon the Newtonian forces are intimately connected with his Third Law of Motion (action = reaction).

[9] So-called "non-holonomic constraints" such as rolling utilize differentials in their articulation, but not in the same way as the $F = ma$-based rules. Hertz stresses the importance of such cases (Hertz 1894, 19–21).

[10] Well, accurately *modulo* the use of analogy! Dissatisfied readers can find sterling (if less vivid) explications of Lagrange multiplier technique in any competent text on differential equations.

problem without engaging in a lot of philosophical grumbling, I wouldn't have much of a future in physics. In this supposition, he was undoubtedly correct; our descriptive practices would become horribly rococo if we seriously attempted to hew to the purified axiomatics that Hertz ultimately recommends.

We've already sketched the serious obstruction that the laissez-faire tolerance of both types of "force" creates: their discrepancies block the derivation of the conservation of energy that Hertz's teacher Hermann Helmholtz originally supplied in 1847. The conservation doctrine maintains that the sum of an isolated system's kinetic and potential energies must remain forever constant. But what, exactly, is a "potential energy"? In Helmholtz's treatment, the term derives from the gauge invariant scalar function V from which the vectoral action-at-a-distance forces acting within the system can be conveniently calculated through partial differentiation (viz., $F_x = -\partial V/\partial x$). As such, the notion had been skillfully employed by Laplace and George Green and is also central in Hamilton's treatments. But for these potential functions to exist, the forces derived from the V can only depend upon the relative displacements between the affected particles; their strengths cannot prove sensitive to their relative velocities as well. Indeed, this requirement explains why Newton's "action = reaction" was reformulated in the "strong Third Law" manner that we utilized in the "Euler's recipe" of the previous chapter. But let's now evaluate a standard constraint force from this perspective, recalling the considerations rehearsed in section (ii). In particular, consider the reaction force F_W that the wire W exerts to hold a sliding bead p in proper position. As we noted, F_W must operate in a *velocity-sensitive* manner, because the faster the bead p travels, a greater degree of acceleration is required to pull it down to its rightful place along the constraining wire W. Accordingly, the wire must be able to "see" the bead's velocity in order to exert the requisite degree of attractive force to pull the bead towards it. This creates a serious foundational obstacle. As long as the status of conservation of energy remains unresolved in this derivational limbo, we lack sufficient "inductive warrant" to resolve our questions with respect to transverse waves in a definitive way.

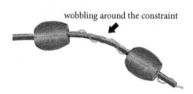

wobbling around the constraint

Because of the "quantum physics amnesia" discussed in Chapter 3, modern commentors are inclined to scoff, "Oh, the proper corrective is readily apparent. We should simply drop all of the constraint forces in terms of an ontology of action-at-a-distance forces acting between point masses." From this perspective, the velocity-sensitive demands of standard constraint forces can be rejected by simply insisting that any seeming allegiance to a higher-scale constraint merely represents a convenient approximation to a more complicated action-at-a-distance reality operating upon the molecular level. Although it may seem as if our bead slides smoothly along its wire, this is an illusion, because the underlying action-at-a-distance binding forces can only maintain the bead's molecules within a close neighborhood of the molecules that comprise the wire. The point masses that properly constitute the bead and wire will in fact wobble about one another in very complicated trajectories that only appear like smooth sliding if we don't look closely enough. From this perspective, all "constraint forces" are merely approximative and are never perfectly implemented within nature. As noted earlier, this is essentially the view of nature that R. J. Boscovich advocated in 1757. But as we also documented in Chapter 2, section (iii), this scheme had become almost universally rejected by Hertz's time as excessively speculative and unable to explain the behaviors of atomic spectra. As a result, Hertz did not include this outdated viewpoint among the "three images" of mechanics that he explicitly considers in his preface.

Indeed, as we also noted in our earlier survey of Euler's recipe, articulating action-at-a-distance laws capable of binding point masses together as stable ensembles is an extremely challenging task that can be avoided by granting the "particles" of nature extended geometries, whether they prove inherently rigid or flexible.[11] In truth, only inherently quantum processes such as the Pauli Exclusion Principle outfit a semi-classical point mass picture of atomic reality with any measure of descriptive validity.

In an attempt to set energy conservation on a firmer basis, Helmholtz later explored foundational schemes that utilize "kinetic + potential energy" variational principles in Hamilton's general manner. This approach comprises the second "image" that Hertz considers and rejects as inadequate in his preface (for the reasons outlined in Chapter 2's appendix).[12] Through this process of elimination, Hertz finally turns to his last possibility, which comprises the mechanism-based picture I have outlined. What I have not sufficiently stressed is how implausible and unattractive this alternative has proved to be, demanding the hypothesization of preposterous ranks of hidden machinery (for details, consult the appendix to Chapter 2).[13] But such is the destination to which we are steadfastly led if we adopt Hertz's policy of purging mechanics of contradictory strands until we reach an

---

[11] The Noll-Truesdell scheme for continuum mechanics surveyed in the appendix below qualifies as such a "force"-centered system (if we generously characterize a stress as a "force"). But it utilizes a large number of conceptual tools that Hertz and his contemporaries did not anticipate.

[12] This is the "energetic image" whose deficiencies for Hertz are outlined in the appendix. The most important of these reflect the subtle difficulties posed by the "non-holonomic constraints" mentioned above. Hertz himself utilizes a somewhat different variational principle borrowed from Gauss as the "dynamical law" he eventually favors (Lützen 2005). I might further remark that in citing Hamilton's principle, we must distinguish between interactive potentials that arise between paired sets of sources V (q_a, q_b) and those that solely upon an absolute configuration $V(q)$. A conventional Third Law point of view demands that the latter be treated as merely approximative, appearing in situations where the q_b source is massive and hard to wiggle.

[13] Hertz's developed proposals rely upon some striking theorems with respect to so-called "cyclic coordinates" as outlined in the appendix to the previous chapter. This otherwise forgotten theoretical alternative played a significant historical role in supplying Pierre Duhem with his chief illustration of the so-called "underdetermination of theory" thesis, viz., for any scientific theory T, there is always a

axiomatizable consistent hull. Surely a better resolution to Hertz's woes can be found that does not drag mechanical practice down the rabbit hole that opens out into the strange Wonderland of his hidden flywheels and linkages. The "architectures" we shall now examine exemplify a more agreeable manner of relief.

APPENDIX

Hertz's Critique of the Third Law

The foregoing exposition highlights the conceptual disharmonies between Newtonian and constraint-style forces in a direct fashion. As such, our discussion adequately illustrates the characteristic manner in which conceptual puzzles can quietly creep into a descriptive practice that improves its practical capacities through gradual emendations. But allied conceptual tensions had burrowed their way into the lore of classical physics long before the conservation of energy (which Newton himself would have rejected) entered the scene. In the preface to *Principles*, Hertz offers a somewhat deeper critique of classical assumption than just outlined. His objections instead center upon some subtle elisions connected with Newton's Third Law of Motion ("action" = "reaction"). As such, Hertz correctly surmised that some of his readers would be suspicious of his critique:

> Now, at first sight, any doubt as to the logical permissibility of [standard mechanical practice] may seem very far-fetched. It seems almost inconceivable that we should find logical imperfections in a system which has been thoroughly and repeatedly considered by many of the ablest intellects. But before we abandon the investigation on this account, we should do well to inquire whether the system has always given satisfaction to these able intellects. It is really wonderful how easy it is to attach to the fundamental laws considerations which are quite in accordance with the usual modes of expression in mechanics, and which yet are an undoubted hindrance to clear thinking.   (Hertz 1894, 5)[14]

Indeed, his worries were well founded, because even his most sympathetic readers have often rejected his analysis. This is partially Hertz's own fault, for the example he favors contains a significant red herring that invites an easy dismissal of the difficulties he raises.

---

distinct rival T* from which it cannot be discriminated upon solely empirical grounds (Duhem 1903, 83–8). This startling claim was beloved of Carnap and Quine and remains popular within philosophy of science circles to this day, despite a paucity of convincing examples. Hertz's force-avoiding construction remains of interest in this role as a seemingly viable competitor to regular mechanics. These topics are not especially germane to our current discussion, however, so I will refer interested readers to Wilson (2017, 196–200).

[14] Hertz regards organizational patterns based upon different selections of materials drawn from (i)–(iv) as different "images" of classical modeling procedure. I will avoid this terminology here. "Logical permissibility" is his preferred term for "logical coherence."

Hertz focuses upon the circumstances of a rock being twirled at the end of a string, and his phraseology suggests that he is merely worried about some familiar confusions with respect to the notion of "centrifugal force." But long before Hertz wrote, mechanical orthodoxy had decided that the latter is not a proper "force" at all, but merely an accelerative pseudo-force.[15] If Hertz's criticism of force-based mechanics merely rests upon this familiar ambiguity, it can be readily dismissed as inconsequential.[16]

As a corrective, let us consider a simpler example provided by Newton himself in which these pseudo-force ingredients are absent, viz., a horse pulling on a large stone with a cord. Newton's Third Law ("action = reaction") somehow contends that certain "forces" must arise in opposed and balanced pairs (F and -F). But where should we expect that these matchups will appear in a given mechanical application? Four central exemplars suggest themselves (illustrated):

(Case i) Forces that arise upon the outer surfaces of two bodies A and B lying in direct contact with another. Newton cites the collision behaviors of billiard balls as direct experimental verification of his Third Law, although in reality his reasoning suppresses the short-term complications of wave movements within the balls (a topic to which we'll later return).

(i) external surface contact   (ii) internal traction forces   (iii) action at a distance forces   (iv) forces at the ends of a connecting rod

A better argument emphasizes the fact that contacting solids can often obtain a state of balanced equilibrium, in which neither object can further penetrate into the space of the other. It is presumed that a balanced contact force will arise across the A/B interface and A's abilities to affect B further must trace to body forces of a non-contact character (e.g., gravitational or electrical attractions).

(Case ii) If we run an imaginary plane P through the interior of a single extended body A, the left-hand portion A_L will push or pull upon its righthand counterpart A_R with a certain degree of internal force F and that A_R will reciprocally do the same to A_L with the force -F (these F's are called "traction forces"). Dissections of this interior character are called "Eulerian cuts" today, and their features are discussed further in the appendix to Chapter 2. If these traction forces Fs act perpendicularly across any plane P we select, this condition is dubbed the

[15] The contrary assumption is occasionally encouraged by some of Newton's own remarks, and a few commentators muddled the Third Law and d'Alembert's principle together. Thomson and Tait (1912, I 240-1) supply a well-known exemplar of this propensity, through recasting a system's inertial behavior as an "action = reaction" response to accelerated motion. This is not the issue with which Hertz is actually concerned. I might further mention that an additional response symmetry known as "material frame indifference" sometimes induces further confusions in these topics as well.

[16] Mach (1896) and Sommerfeld (1943) interpret this passage in this unflattering manner, although otherwise their readings of *Principles* are generally sound.

local "internal pressure" operating within A. If the F's instead match up along a skewed angle, the condition is called an "internal stress."[17] The historical development of these two terms ("internal pressure" and "stress") is complicated and confused (Truesdell 1968), with many authors (including Euler himself) attempting to surreptitiously reduce such actions to surface contacts between contacting "molecules" (case i) or action-at-a-distance relationships between separated small bodies (case iii). Hertz himself does not acknowledge this second species (ii) of Third Law balance either, for reasons that we'll consider in a moment. In truth, coherent notions of both "stress" and "internal pressure" only make sense within a continuum mechanics framework, which is a development that neither Hertz nor most of his Newtonian opponents actively considered. I'll return to this datum in a moment.

(Case iii) If a body A acts upon body B with a force F across an intervening spatial gap, then B will likewise act on A with the action-at-a-distance force -F. In point mass physics (as codified within the "Euler's recipe" of the previous chapter), this is the only category of Third Law balance that appears, simply because unextended points can't lie in meaningful "surface contact" with one another.

(Case iv) At the two ends of a rigid rod or taut rope that connects A with B. This application is commonly invoked in practice, but its proper rationale comprises the locus of the concerns we will now examine.

With respect to (i), (iii), and (iv), Newton appears to have presumed that some form of deductive thread connects these three demands, although his exact policy for doing so is hard to decipher (Wilson 2020a). In his penetrating survey *Physics for Mathematicians*, Michael Spivak pungently comments upon the "audacious generalizations" involved:

> Strenuous exertions have allowed us to tease out a bit of meaning from the first two laws, both of which involve individual bodies, but say nothing about the interactions between different bodies. This information is given by Newton's third law, and since all of mechanics supposedly rests on Newton's three laws, this one must really be a doozy. In fact, it is usually stated as a memorable apothegm: "Every action has an equal and opposite reaction". In this form it is ideally suited to misappropriation by armchair philosophers, moral and political thinkers, and others of that ilk.... Newton took a basic experimental fact, the conservation of momentum in collisions, and recast it as a corollary of an apparently equivalent formulation, Newton's Third Law. Nowadays, physics texts may allude to the essential role played by the third law— "forces always appear in pairs"—but they give scant attention to the fact that Newton's decision to cast those experimental results in terms of the third law was an incredibly audacious generalization! Based on results involving the completely unknown repulsive forces between colliding bodies, Newton hypothesized a more general law concerning all forces.... Though it might seem that we have provided far too extensive a discussion of the third law, which merits no comparable attention in physics books, there is an equal and opposite reaction, that physics books provide far too little discussion. One might well wonder why critical readers readily accept so general a law buttressed by so little experimental evidence, as if it somehow expresses a morally compelling symmetry. Perhaps it's just because it's so easy to confuse laws of nature expressing the symmetry of space with other laws that merely seem to.
>
> (Spivak 2010, 21–5)[18]

Spivak is right to characterize our complete Third Law package as "really...a doozy," and we shall shortly supply a Hertzian diagnosis of its partially illicit nature. But let us first

---

[17] Actually, these tractions must be force *densities* as explicated in the appendix to Chapter 2.

[18] In Wilson (2020a), I explain the dodgy Noether's theorem considerations to which Spivak alludes.

scrutinize our horse, rock, and rope more closely. If we introduce several Eulerian cuts into the rope, e.g., at positions A and B, our internal pressure principle (ii) dictates that locally these pressures must lie in F/-F balance across the cuts. So far, so good. But does any sound consideration further assure us that these local interior

forces must be identical at the distinct locations A and B? Not always; a secret appeal to rope equilibrium has crept into our deliberations.

To see this, let's first consider a presentation of the Third Law that invokes our (case iv) situation in a naïve way. William Whewell offers the following in an old textbook of his:

> *A Pressure transmitted directly* in the straight line in which it acts, is transmitted without augmentation or diminution, whether by means of a rigid or a flexible body. If we push by means of a rod, if we push by means of a cord, the force transmitted to the farther extremity of the rod or the cord is exactly the same as that which we exert at the end which we hold. In stating this, we of course leave out of consideration the weight of the rod and of the cord.
>
> That matter thus transmits force directly, without altering its amount, is a necessary consequence of this equality of action and reaction. When a force which acts at A in the direction of the straight line AB, transmits a force to the point B, it may be counteracted by an opposite force at B, acting in the direction BA. These opposite forces at A and at B which thus balance each other must necessarily be equal; for each may be considered as the reaction with reference to the other.   (Whewell 1847, 10)

Whewell is scarcely idiosyncratic in advancing this claim, for one can readily find similar conclusions drawn without apology within most surveys of mechanics, ranging from Newton's own writings to modern-day texts.

But Whewell's casual appeal to "transmission" along the rope shows that some apology is needed because a constant F/-F force balance along the entire rope can obtain only if the latter has reached a condition of static equilibrium. Consider what happens when the horse first pulls on its end of the rope. Plainly the rock will not immediately discern an equal reaction pull at its faraway end, for that portion of the rope may remain limp until a pressure wave reaches it. Indeed, even after this first signaling occurs, a further expanse of relaxation time must transpire before the pressures everywhere along the rope stabilize upon some constant steady state value. Indeed, if the horse proves feeble or the rope flexes like a limp noodle when it is pulled, equilibrium equal pressures may never be reached. In addition, a substantial number of frictional factors are required to smooth out the traveling waves directly originating at the horse into an eventual state of constant tension. From a pressure-centric point of view, the horse's original pulling on the rope merely initiates a complex *causal process* that only later eventuates in an equilibrium condition in which the horse's force upon the rock (F_H) equals the negative of the rock's direct pull on its end ($-F_R$). But Whewell's form of the "Third Law" reads as if none of these intermediary smoothing events are obligatory.

As we observed in the appendix to Chapter 2, the (ii) notion of "internal pressure" is a rather sophisticated construct that is most conveniently understood in terms of the Euler cut principle central to continuum mechanics. By placing several of these cuts inside the rope at positions A and B, we will find the tractions across each of them will be equal and opposite in agreement with our Third Law (ii) reading, but that there's no reason to suppose that the forces at A bear any resemblance to those at B until some complicated process brings the entire rope into stasis.

But a true "first image" mechanics cannot appeal to internal continuum behavior in this direct way and must devise some complicated molecular mechanism to induce a suitable simulacrum of rope stabilization solely in terms of molecule contact forces (i) and action-at-a-distance forces (iii). But this is a very tall order, involving a large degree of speculative (and implausible) molecular hypotheses. Clearly, if we wish to found classical mechanics upon Hertz's force-based "first image," we are not allowed to freely borrow conceptions like "internal pressure" from the completely different "image" of continuum mechanics.[19]

But let us now replace the rope between the horse and rock with a rigid rod in Hertz's favored "third image" manner, in which case the posited rigidity will guarantee that F_H will instantly equal $-F_R$, because as Descartes observed:

> I shall observe that while a stone cannot pass to another place in one and the same moment, because it is a body, yet a force similar to that which moves the stone is communicated exactly instantaneously as it passes unencumbered from one object to another.... For instance, if I move one end of a stick of whatever length, I easily understand that the power by which that part of the stick is moved necessarily moves also all its other parts at the same moment, because the force passes unencumbered and is not imprisoned in any body, e.g. a stone, which bears it along.
>
> (Descartes 1968, 33)

So if we accept totally rigid bodies within our mechanics in Hertz's favored manner, requirements (i) and (iv) will collapse into one another and needn't be distinguished. The rod will then act in the manner of a constraint, and its constant length serves as the cause of why F_H has to equal $-F_R$. At the cost of a minimal terminological substitution,[20] we can then capture Hertz's original critique as follows:

> Can we, without destroying the clearness of our conceptions, take the effect of [length] twice into account,—firstly as [length], secondly as force? In our [first image] laws of motion, force was a cause of [length], and was present before the [final length stabilized]. Can we, without confusing our ideas, suddenly begin to speak of forces which arise because of [a constant length], as a consequence of [that length]? Can we behave as if we had already asserted anything about forces of this new kind in our laws, as if by calling them forces we could invest them with the properties of forces?
>
> (Hertz 1894, 6)

---

[19] In Chapter 2, I explained why Hertz does not explore this continuum mechanics image as one of his alternatives.

[20] Hertz articulates his critique in terms of an unfortunately articulated gloss on the state that obtains when a rotating rock on a string reaches steady state equilibrium:

> With regard to this opposing force the usual explanation is that the stone reacts upon the hand in consequence of centrifugal force, and that this centrifugal force is in fact exactly equal and opposite to that which we exert. Now is this mode of expression permissible? Is what we call centrifugal force anything else than the inertia of the stone?   (Hertz 1894, 6)

But this "usual explanation" incorporates the force/pseudoforce confusion which any sensible Newtonian will surely reject. I believe that he should have written:

> The usual explanation is that the rope acts upon the hand in consequence of its reversed equality with the force that the rope acts upon the rock, which must exert a force of exactly the right magnitude F_R to pull the rock from its inertial path to match the required length of the taut rope. But F_R's required magnitude can be immediately calculated from the rock's mass and current velocity without any assistance from a Newtonian-style force law.

Note that knowledge of velocity is required for this calculation, contrary to the "strong reading" of the Third Law that we required in the Euler's recipe of Chapter 2.

In mathematical terms, we again confront the contrast between an equation of constraint (which is completely independent of time in the case of a rigid rod) and evolutionary differential equations of an $F = ma$ type which must be integrated to obtain quantities such as lengths.[21] From a constraint-centered point of view, the constant length of the taut rope represents the true "cause" of the fact that the horse's force upon the rock (F_H) becomes immediately transferred to the rock on the other end -F_R.[22] These subtle adjustments in the application of "cause" will form the primary topic of Chapter 7.

In short, what we have discovered is that standard mechanical practice tacitly borrows assumptions from divergent foundational formats to which none of them is fully entitled. Hertz's force-based "first image" can legitimately demand tenets (i) and (iii), but not (iv), whereas his own "third image" approach can accept (i) and (iv) but should reject (iii) as inadmissible. For the reasons surveyed in Chapter 2, Hertz does not include the most plausible resolvent of these difficulties in his inventory (the localized Euler's cut pressure balance (ii)) because of its intrinsically "continuum physics" conceptual requirements. This omission makes his reasoning somewhat hard to follow.

In any case, we can now see that these Third Law/equilibrium ambiguities equally apply within Hertz's original example, in which our stone is placed at the end of a rotating rope. By investigating the static circumstances of our horse and load instead, it becomes evident that the centripetal pseudo-forces involved in the case of the rotating rock do not play a significant role in the applicational ambiguities at the center of Hertz's penetrating diagnosis. Once again, this irrelevant intrusion has frequently led to a trivialization of his true insights.

As a practical matter, mechanical practice has thrived precisely because it freely invokes any of our four Third Law readings as applicational circumstances suggest, in spite of these underlying inconsistencies. Indeed, my Physics 101 instructor can be pardoned for attempting to inculcate the free-wheeling liberalities of established practice within my stubborn teenaged psyche. But is there some other escape route whereby we might rationalize our everyday "use all four" policies in a less brutal manner than the axiomatic dismemberment that Hertz recommends? This will be the chief topic of our next chapter.

[21] More exactly, standard mechanics utilizes Lagrange multipliers to convert these constraint demands into terms that can be sensibly comingled with the force-based factors The former then become dubbed "constraint forces" as a result of this projection into the force-centered realm (these are the circumstances that I've attempted to explicate within the military metaphor of section (iii)). In a purely force-based setting such as celestial mechanics, the Euler's recipe differential equations will completely determine the system's subsequent motions. But when constraints on movement are added to the picture, the "causal" determinations flow in the opposite direction: the constraints make the differential equations behave differently than they otherwise would. This represents the central foundational ambiguity that Hertz correctly identifies. These remarks represent a truism within the modern theory of differential equations encompassing constraints.

[22] Of course, Hertz is obliged to explain why real-life ropes sometimes become limp. To do so, he must posit that they are secretly comprised of mechanical linkages of a preposterous complexity.

5

Multiscalar Architectures

> If we study the history of science, we see happen two inverse phenomena.... Sometimes simplicity hides under complex appearances; sometimes it is the simplicity which is apparent, and which disguises extremely complicated realities.... No doubt, if our means of investigation should become more and more penetrating, we should discover the simple under the complex, then the complex under the simple, then again the simple under the complex, and so on, without our being able to foresee what will be the last term. We must stop somewhere, and that science may be possible, we must stop when we have found simplicity. This is the only ground on which we can rear the edifice of our generalizations.
>
> Henri Poincaré (Poincaré 1905, 105–6)

(i)

In the previous chapter, we compared standard physical practice to an Escher-like staircase in which every time we fancy that we have reached a lowest landing, we find an inferior level lying just ahead. Following this continuously downward path, we eventually come back to where we started, at least with respect to the descriptive mathematics we employ to capture the behaviors of nature (whether it be the mathematics of point masses, of rigid mechanisms, or of truly flexible continua). In his attempts to rectify these twisty circumstances, Hertz lands himself within an exotic landscape of mechanism-patterned media whose behaviors cannot be seriously regarded as corresponding to nature in any plausible way.

So much for the righteous pathways of rigorous purification. But perhaps we needn't insist upon a single-leveled axiomatization of this sort. Perhaps we should only hope that our tools of mathematical description will prove adequate to the *dominant physical behaviors* that we normally witness upon a particular scale length, leaving further investigations into their currently suppressed details to other occasions? The nineteenth-century physicist P. G. Tait develops this theme as follows:

> [A]ll who have even a slight acquaintance with [Newtonian physics] know that the laws of motion, and the law of gravitation, contain absolutely all of Physical

Imitation of Rigor: An Alternative History of Analytic Philosophy. Mark Wilson, Oxford University Press. © Mark Wilson 2022.
DOI: 10.1093/oso/9780192896469.003.0005

Astronomy, in the sense in which that term is commonly employed:—viz., the investigation of the motions and mutual perturbations of a number of masses (usually treated as mere points, or at least as rigid bodies) forming any system whatever of sun, planets, and satellites. But, as soon as physical science points out that we must take account of the plasticity and elasticity of each mass of such a system, the amount of liquid on its surface, ... [etc.], the simplicity of the data of the mathematical problem is gone; and physical astronomy, except in its grander outlines, becomes as much confused as any other branch of science.

(Tait 1895, 9–10)

Of course, when Tait speaks of treating a good-sized planet "as a point," he merely means that its dominant behaviors can be descriptively captured with three degrees of freedom, employing some convenient fiducial point to stand in for the full body. Nor need we assume that this descriptive system can permanently remain mathematically adequate unto itself, for potential collisions, earthquakes, and other sorts of anomalous calamities may intervene. At such a time (and for the many other reasons that Tait outlines), we will need to "open up the suppressed degrees of freedom" and consider revised modelings in which our planets are allowed to flex a bit or where we consider the frictional effects of placing a liquid ocean on their surfaces. In the jargon we shall employ throughout this chapter, point mass mechanics captures the *dominant behaviors* we witness in "Physical Astronomy" (= the conventional theory of the solar system), but we must frequently adjust our focus to consider effects that arise upon smaller-size scales:

[A]s long as we are content to view solids as perfectly smooth and rigid, and fluids as incompressible and frictionless, the difficulties of Dynamics, though often enormously great, are entirely mathematical, it falls naturally into quite distinct and separate head, and the classification of its various problems is comparatively simple. Introduce ideas of strain and fluid friction, ... and the confusion due to imperfect or impossible classification comes in at once. Each problem, instead of being treated by itself,

Tait

has to borrow sometimes over and over again, from others; and the only fully satisfactory and uncomplicated mode of attacking such a subject (were it conceivable) would be to work it all out at once. Hence in dealing with the general subject of heat we shall find it quite impossible to lay down definite lines of demarcation. Divide it as we choose, each part will be found to require for its development something borrowed from another.   (Tait 1895, 9–10)

Such observations will strike the reader as familiar and comparatively obvious, and some years ago I complained about Tait's chains of deferred corrections as illustrating "the lousy encyclopedia phenomenon":

> after a regrettable "reference work" that my parents had been snookered into purchasing (the 1950s represented a notorious era of encyclopedia mania). As a child, I would eagerly open its glossy pages to some favorite subject ("snakes," say). The information there provided invariantly proved inadequate. However, hope still remained, for at the end of the article a long list of encour-
>
>
>
> aging cross-references was appended: "for more information, see rattlesnake; viper; reptile, oviparous..." etc. Tracking those down, I might glean a few pitiful scraps of information and then encounter yet another cluster of beckoning citations. Oh, the hours I wasted chasing those informational teasers, never managing to learn much about snakes at all!   (Wilson 2006, 180)

(ii)

What neither I nor Tait realized at that time is that linked descriptive packets of this potentially cyclic character can not only prove informative; they can facilitate a virtually complete mastery of a complicated set of behaviors at relatively low computational cost.[1] Indeed, this simple prospect comprises the underlying basis of many significant advances within modern computing, often collectively labeled as "the multiscalar revolution." The basic trick is to position a variety of individual modeling tactics (called "submodels") within a coordinated architecture that shifts between these components in a controlled, checks-and-balances manner. By doing so, a multiscale plotting can resolve the computational hazards that Terrance Tao calls the "curse of dimensionality":[2] keeping track at one time of all the interactive variables relevant to a complex system will easily swamp the capacities of the most compendious of imaginable computers:

> The inability to perform feasible computations on a system with many interact-ing components is known as *the curse of dimensionality*. Despite this curse, a

---

[1] An allied insight is contained within Anil Gupta's approach to "circular definitions" (Gupta 2011, 73–134). A sequence of definitions may run in an unfounded circle, yet their collective entanglements may supply extremely valuable information along the way. Example: "i = the number whose square = –1 and is further identical with i itself" circularly contains its *definiendum* within its *defiens* but nonetheless tells us everything we need to know with respect to the arithmetic of complex numbers. Indeed, real-life reference works frequently exemplify "the lousy encyclopedia phenomenon" to excellent purposes.

[2] "Tyranny of scales" also serves as an evocative label for the phenomenon (Batterman 2013).

remarkable phenomenon often occurs once the number of components becomes large enough:... the aggregate properties of the system can mysteriously become predictable again, governed by simple laws of nature. Even more surprising, these macroscopic laws for the overall system are often largely *independent* of their microscopic counterparts that govern individual components of the system.

(Tao 2012, 24)

For clarity (if not eloquence) we might re-express Tao's concern as "the curse of many degrees of freedom," because it directly captures the fact that the raw number of variables we may need to consult within a complex material typically prove unmanageably huge. The "simplicity" to which Tao appeals derives from the convenient fact that these variables can often be partitioned into smaller "dominant behavior" collections associated with characteristic scale lengths ΔL. This is because the prevailing activities witnessed upon that scale frequently obey relatively simple prin-

ciples, at least with respect to run-of-the-mill events (which represent the "dominant behaviors" in question). For example, if we shear a slim polycrystalline material such as a rod of hard plastic with an applied force **F**, its component crystal units will usually remain firmly attached to one another along their interfacial

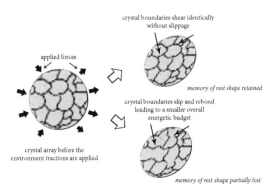

crystal boundaries shear identically without slippage

applied forces

memory of rest shape retained

crystal boundaries slip and rebond leading to a smaller overall energetic budget

crystal array before the environment tractions are applied

memory of rest shape partially lost

boundaries. As long as this behavioral presumption remains true, the rod will act upon a macroscopic-scale length Δ_H like an elastic spring, in which the specimen "remembers" its pre-stressed condition and returns to that original shape once the applied pressure **F** is removed. If so, the rod will obey the simple Hookean rule that $\Delta L = kF$ on the macroscopic level, where ΔL represents the difference between the rod's rest length and its extended or compressed displacement. However, if the applied force **F** becomes too extreme, the tiny crystal components within the rod may slip along their interfacial boundaries upon a microscopic level Δ_L and rebind to one another at altered positions. If this happens, so-called "plastic flow" will have damaged our rod, causing it to lose some of its previous "elastic" capacity for remembering its original shape perfectly. Approached at the macroscopic scale Δ_H of the rod itself, satisfactory rules for plastic flow are hard to formulate, but if we shift our attention to the microscopic level Δ_L, nice models for the boundary reattachment can be developed in terms of relatively simple considerations of energetic allocation. We should ask, Can this material reduce its total allotment of trapped internal energy by reconfiguring the chemical bonds that tie its component grains together? (Familiar example: a highly stressed egg will fracture into pieces to prevent the applied strains from promulgating throughout the entire shell). Frequently a simple threshold test decides the result: as long as the induced stresses upon the crystals remains low, the energetic cost of adjusting its grain boundaries will be too high. But if the applied stresses become greater, it may be better to devote some energy to overcoming the barriers against grain alteration in order to free the crystals inside from a state of extreme elastic shearing. In other words, imitate the noble egg and crack apart rather endure a high degree of imposed bending (an allied analogy: a house blows apart in a hurricane once the energetic cost of resisting the winds exceeds the price of extracting its nails). So the behavioral rule pertinent to interfacial realignment within a crystal laminate upon the Δ_L level represents a fairly simple question of a chemical threshold, determined by comparing the cost of maintaining a high degree of internal strain energy within the grains against the cost of adjusting their interfacial connections. The situation nicely illustrates Tao's distinctions: most of the time the "macroscopic laws for the overall system" (in the situation at hand, Hooke's behavioral rule for the rod formulated at the Δ_H level) remains "largely independent of its microscopic counterparts" (viz., how its component crystals behave at Δ_L). When this Δ_H level "dominant behavior" presumption fails (as it inevitably will), we can still locate tractable forms of behavior (viz., governing the normal processes of boundary reformation) at some appropriate lower-scale Δ_L. Obedient to the adage "don't scratch where nothing itches," we should inspect our composite material upon the lower ΔL level only when some warning consideration suggests that our usual Hookean rules may potentially fail upon the Δ_H level. As anyone who deals with a neurotic fussbudget knows, we can save enormous amounts of time if we refrain from fretting over every unlikely eventuality.

The general observation to which Tao's remarks lead is this. If we can isolate a suitable hierarchy of simply characterizable dominant behaviors arising upon a succession of characteristic length scales, we can conquer his "curse of dimen-sionality" by linking these individual modeling opportunities together into a mutually re-enforcing computational network as "submodels." The guiding strat-egy behind this computational policy recommends that we should remain content with our higher ΔL scale "dominant behavior" calculations until some warning criterion instructs us to consult a submodel associated with the scale length ΔL^* as a test of whether our "dominant behavior" presumptions at the ΔL scale remain valid. If not, the ΔL^* submodel will need to send a corrective message back to the original ΔL submodel in a manner we shall discuss shortly. And we may still need to inspect further submodels ΔL^{**}, $\Delta L^{***}, \ldots$ for similar intrusions of irregular damage. Reverting to my "lousy encyclopedia" analogy, we stick with the entry on

"Snake" until some consideration moves us onto "Reptile." And then we should continue searching through further volumes until we feel confident in the scattered data we have acquired. More exactly:

> Probe the intricacies of a complex problem only to the depth required to extract larger-scale answers that are normally trustworthy. Simultaneously conduct smaller-scale tests to verify that these presumptions of "normalcy" are actually satisfied.

The computational efficiencies to be gained by simply not bothering with a material's finer-grained behaviors if we already know the higher-scale answers are truly stupendous. Such tactics have reduced the computer time required to simulate a complex material reliably by many orders of magnitude (I'll supply a simple illustration of such a scheme in section (iv)).

(iii)

However, to better appreciate how these advantages operate, let us first consider several manners in which we reason fruitfully about simple target physical systems in situations where multiscalar tactics aren't yet required. Perhaps my chief objection with respect to standard theory T characterization is that its schematisms collapse distinct forms of explanatory pattern into misleading commonality. Indeed, in one of his rare departures from descriptive accuracy, Hertz does exactly this in what is possibly the most widely quoted passage from the *Principles*:

> We form for ourselves images or symbols of external objects; and the form which we give them is such that the necessary consequences of the images in thought are always the images of the necessary consequences in nature of the things pictured.... [W]e can then in a short time develop by means of them, as by means of models, the consequences which in the real world only arise in a comparatively long time, or as a result of our own imposition.
>
> (Hertz 1894, 1)

This citation suggests that successful modeling should imitate the unfolding stages of a target physical process in the manner in which Harpo mimicked Groucho in *Duck Soup*. With a fair measure of forbearance, we can concede that certain forms of successful modeling approximately resemble this recipe in containing computational steps that parallel physical developments to a reasonable degree. For example, we might track the developing path of a cannon ball

by utilizing a simple computational procedure called "Euler's method" in the ball's
forward motion is successively plotted
over increasing units of time Δt by progressively filling in small squares on a piece of
graph paper.[3] Techniques of this sort (of
which more effective versions such as
Runge-Kutta methods are available) are
called "marching methods" simply because
they tramp across the graph paper in approximate fulfillment of the manner in which
Hertz's unfolding progressions of "ideas" are
expected to mimic the unfolding stages
within our cannonball's flight.

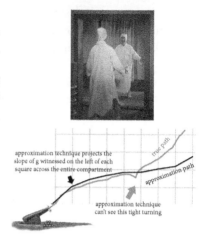

approximation technique projects the
slope of g witnessed on the left of each
square across the entire compartment

true path

approximation path

approximation technique
can't see this tight turning

But this presumed parallelism can prove approximate at best, because the
potential field g through which our ball flies may contain tiny blips which our
crude Δt-width approximations are likely to overlook completely (the rule only
pays attention to g's value at the start of each Δx interval). But such a tiny
irregularity within g may divert our cannonball in some entirely novel direction,
which our Eulerian calculation will completely overlook. As a result, our calculations will merrily transport the projectile along a completely erroneous path. Our
only hope of completely avoiding this source of error is to reduce our "step size"
Δt to 0, which means that we're no longer dealing with a reasoning technique at
all, but simply a differential equation solution whose unfolding behavior may
remain entirely unknown to us.

For later purposes, let us now consider an equally popular reasoning technique
that may seem superficially similar to a Hertzian marching method but actually
relies upon a significantly different explanatory basis. Let us load an elastic
clothesline with garments and attempt to predict how the rope will sag. To

---

[3] In particular, assume that our ball obeys the simple rule $mdy/dx = g(x)$ where $g(x)$ represents a
variable gravitational potential of some kind. Euler's marching method calculations presume that the
slope of our ball's trajectory at the start of each "step size" Δx interval will nearly remain across the rest
of the interval (at which point the slope is adjusted according to the rule $g(x + \Delta x)/m$). More
sophisticated refinements (such as the popular Runge-Kutta methods) improve the accuracy of these
approximations by various backtracking devices. "Step size" schemes of this general Δx character
usually emerge as "finite difference" approximations to an originating collection of differential equations. Many of the logic-based distortions characteristic of "theory T thinking" stem from treating these
finite difference replacements as if they are somehow "equivalent" to their differential equation
progenitors (we will confront several exemplars of this faulty equivocation in Chapter 6 when followers
of Hume subvert a properly continuous causal process into the "cause at t, effect at t + Δt" approximations favored within Hume's own discussion). I've tried to minimize direct consideration of these
unfortunate forms of logistical surrogate within this book, but sometimes the misrepresentations of
explanatory pattern encouraged by these replacements (e.g., Carl Hempel's well-known "D-N model
of explanation) demand more overt attention. Wilson (2017, 51–98) contains a fuller examination of
these issues.

support a piece of clothing of mass m_i located at i, the cord must locally bend enough that the projected upward thrust of the rope's tension T_ϕ must equal m_i at i. But how might we determine the magnitude of this tension? Answer: add up all of the stretched lengths of the segments spanning the step size Δx and determine how the stretched cord length L_ϕ compares to its unloaded length L_0. Hooke's law then dictates that the tension produced T_ϕ will equal k $(L_\phi–L_0)$.[4] We can safely assume that our clothesline will adjust its local curvatures until the rope reaches a

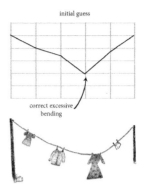

initial guess

correct excessive bending

final configuration ϕ in which its internal tension T_ϕ becomes minimal. But how should we ascertain that finalized value? Suggestion: start with an arbitrary trial proposal ϕ_0 and alter its local curvature at i so that our corrected curve ϕ_1 perfectly supports the heaviest piece of clothing m_i at its point of attachment i (in the jargon, we set its "residual" to 0). Adopt ϕ_1 as a revised guess and now examine the bending at some other loading point j. Even if the local load m_j had been properly supported at j within our original guess ϕ_0, the subsequent adjustment to ϕ_1 may have spoiled this balancing. If so, let's again tweak the local curvature at j to reach a revised trial solution ϕ_2 and continue our examination. And so it goes, until we (hopefully) reach a final proposal ϕ_∞ in which no further adjustment is required (ϕ_∞ is called the "fixed point" of the sequence of ϕ_0, ϕ_1, ϕ_2,...). This form of calculational tactic is often called a "relaxation method" on the grounds that if the clothesline were ever positioned in a non-optimal configuration ϕ^*, it would eventually shed its excess degree of tension by "relaxing" to the equilibrium condition ϕ_∞.

Reasonings that operate in this general manner are also said to proceed by *successive approximations*, and their component steps can be nicely codified within a standard feedback flow diagram. Their basic feature is that our calculations don't escape from their repeated revision demands until a satisfactory "fixed point" is reached. Hundreds of iterations of this type are sometimes required before suitable results become stabilized, which may suggest that this sort of calculation will be rather inefficient. But this impression is erroneous, as we shall see when we examine how allied tactics supply a typical multiscalar setting with its computational advantages.

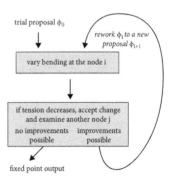

trial proposal ϕ_0

rework ϕ_i to a new proposal ϕ_{i+1}

vary bending at the node i

if tension decreases, accept change and examine another node j
no improvements possible   improvements possible

fixed point output

---

[4] Expressed as a differential equation for a continuous distributed load ρ, these relationships become the familiar $k\partial^2 y/\partial x^2 = \rho$.

However, the label "relaxation method" can prove misleading if we presume that our sequence of ϕ_0, ϕ_1, ϕ_2,... will directly parallel the actual series of cord adjustments ϕ_0, S_1, S_2,... that we will witness if we initially hold the cord in the configuration ϕ_0 and turn it loose. Quite the contrary, no such correspondence with real-life "relaxation" is likely to appear at all. Indeed, a real-life clothesline will never relax into stasis unless some frictional mechanism enters the picture to gradually drain away the rope's kinetic energy. We will need to attend to these intervening frictional events if we want to predict the actual sequence of rope events ϕ_0, S_1, S_2,... that we will actually witness as a loaded clothesline subsides into quiescence. This marked discrepancy between *computational path* and *physical trajectory* nicely illustrates the misleading aspects of "consequences in thought/consequences in nature" metaphor.

Hertz, of course, would have immediately conceded the justice of these complaints had they been suggested to him: "Oh, I was merely attempting to provide my readers with a general idea of how science works." But why invoke a schematic representation of "how science operates" that is so palpably misleading? I'll postpone a fuller discussion of this question until our concluding chapters.

For more immediate purposes, let us observe that the differences in reasoning technique between a marching method and a relaxation method mode of calculation signal a significant divide with respect to their underlying patterns of explanatory appeal. Although applied mathematicians pay careful attention to these methodological differences, philosophers frequently do not, leading to a significant decline in their diagnostic capabilities with respect to scientific argumentation. Mathematicians classify our cannon ball modeling as being of *evolutionary type*, indicating that its applicable formulas temporally advance a target system's development smoothly forward in time. But when they turn to our clothesline example, they characterize it as a *revised equilibrium modeling* that makes no attempt to track the target system's shifting temporal behaviors at all and instead leapfrogs forward to an eventual condition when the system has once again subsided into statis.[5] That is, in an equilibrium modeling framework we forgo all ambitions of stage-by-stage tracking and immediately focus upon the system's eventual condition when it returns to rest. How long will this subsidence into equilibrium require and which frictional factors will allow the system to reach it? An equilibrium modeling doesn't attempt to supply answers. But often those details of intermediary behavior aren't particularly important, and reliable answers are hard to reach (because the calculations are highly sensitive to error). So what? We can be just as certain of the fact that loaded clotheslines eventually come to rest as the datum that our bodies possess heads. So why not rely upon that firmly grounded knowledge and forego further fussing about our clothesline wiggles on its way there?

---

[5] So-called "steady state" modelings display many of the same conveniences as our equilibrium treatments, but I won't expand upon these themes here.

When we look at the details, we'll find that standard multiscalar schemes gain many of their computational efficiencies by relying upon equilibrium submodels whenever they can. And we do the same in our everyday reasoning as well. In the next chapter, we'll find that some of the puzzling behaviors of the word "cause" are closely entangled with these considerations of effective choices of submodel.

In other writings (Wilson 2017), I have complained vigorously (possibly to extremes) of the ways in which standard theory T thinking indiscriminately mashes these distinct modeling categories together, brutalizing along the way the well-honed discriminations that mathematicians have devised for keeping these matters straight (viz., "initial conditions," "boundary conditions"). As the present book is intended as a *vade mecum* with respect to philosophical investigation generally, rather than as a dedicated project within descriptive philosophy of science per se, I'll not belabor these issues further. For our purposes, two vital pieces of methodological consideration will be required. (i) Our reasoning policies often reach useful conclusions through repeated sequences of *successive approximation*. (ii) Advanced *partial knowledge* that a target system will reach an equilibrium condition quickly can be deftly exploited to simplify our reasoning tactics significantly.

(iv)

Let us now examine how modern computer scientists have converted these humble observations into integrated programs that function with amazing effectiveness. I will illustrate these techniques with a few easy-to-illustrate examples, leaving aside the complications that invariably complicate more realistic examples.[6]

Let us set ourselves the following task. Some gambling magnate wants to erect a massive casino above large chunk of buried granite. On the macroscopic scale (which I will symbolize as ΔL_H), granite acts like a simple Hookean material: it compresses linearly and isotropically under pressure governed by two elastic constants (E and μ). Such events can be considered as the

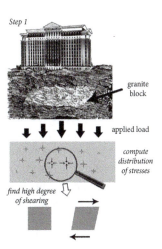

Step 1

granite block

applied load

compute distribution of stresses

find high degree of shearing

---

[6] I apologize for the somewhat unrealistic constructions utilized in this chapter. The potential mechanisms of lower-scale damage included within my specimen architectures have been selected because they are easy to explain, not for their real-life plausibility. Accordingly, I largely ignore thermal factors, for example, despite the fact that they generally serve as major determinants of whether such damage will arise or not (see the brief remarks on "energy cascades" in Chapter 7). Significant probabilistic concerns usually require attention as well, but their inclusion would significantly retard my narrative arc to no significant philosophical end.

rock's *dominant behavior* at the length scale ΔL_H obedient to a simple set of equations originally developed by Charles Navier.[7] But these merely capture the stone's smoothed over behavior upon a lengthy observation scale. In itself, granite represents an extremely complicated structure, consisting in a large number of readily distinguished compositional layers. The fact that its ΔL_H scale behaviors submit to a simple Navier modeling represents an admirable illustration of Tao's observation that "once the number of components becomes large enough:... the aggregate properties of the system can mysteriously become predictable again, governed by simple laws of nature." In the sequel I shall sometimes characterize easy-to-model platforms of this ilk as "descriptive opportunities" (viz., convenient gifts from nature itself, not of our own making).

But no opportunity is ever perfect, and granite, in the final analysis, remains a highly complex material. But if we naïvely attempt to tackle our casino loading problem from the atomic level upwards, our efforts will immediately fall prey to Tao's "curse of dimensionality," for the variable set required would immediately swamp the capacity of any conceivable computer.[8] So let's implement the methodological advice recommended earlier:

> Probe the intricacies of a complex problem only to the depth required to extract larger-scale answers that are normally trustworthy. Simultaneously conduct smaller-scale tests to verify that these presumptions of "normalcy" are actually satisfied.

Here's how we might do that. Step 1: using our ΔL_H-level Navier equations, compute a trial distribution of stresses inside the rock $\mathcal{S}_0$. If the casino loading is complex, this computational task is likely to prove lengthy in itself, requiring finite elements or some allied form of numerical method. As in our clothesline routine, we should scour these results looking for locales where the predicted shearing stress is very high (symbolized by a magnifying glass in the illustration). These represent the regions where the dangers of lower-scale damage are highest, where the local specimen of rock might collapse in a manner that invalidates the Navier rules in its vicinity.

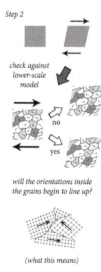

Step 2

check against lower-scale model

no

yes

will the orientations inside the grains begin to line up?

(what this means)

Step 2: To estimate the likelihood of this damage occurring, we frame a newly localized submodel pitched at upon a suitable mesoscale ΔL_L that divides a small region of rock region into a so-called *laminate*: an array of contacting

---

[7] $\sigma_{ji,\,j} + f_i = 0$, with their pertinent compatibility demands.
[8] Such a naïve modeling will further demand vast expanses of starting data that we will never obtain.

mineral grains bearing individualized orientation arrows as shown.[9] This shift represents an adjustment in the mathematical tools utilized; the continuum physics of a laminate containing interfacial boundaries is more complicated than the smoother physics than we first employed upon the original resolution length ΔL_H where a smoother form of governing physics suffices (viz., Navier's equations). These alternations in submodel are required because the composite crystals within granite are individually *anisotropic* in their local responses to pushes and pulls, indicating that these tiny

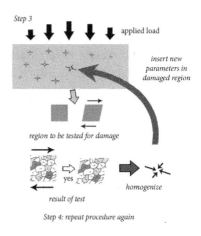

portions of rock grain will shear to different degrees depending upon whether the local load that trickles down from the casino acts along the grain of the arrows or against them (wood provides a familiar illustration of a strongly anisotropic material). As a granular composite, isotropic granite runs a danger of becoming metamorphized into anisotropic gneiss if these grain arrows, which are randomly distributed within a normal piece of granite, begin to partially align under prolonged stress with their next-door neighbors. To be sure, this transmutation rarely occurs in granite under normal conditions because the grains must internally reconstitute themselves in a manner that demands a large influx of external energy (so we confront a behavioral threshold similar to that pertinent to the plastic flow discussed above, although we are now considering a somewhat different form of microscopic damage). In our stage 2 test, we will consult our composite physics submodel to determine if the specimen's localized loading exceeds this energetic threshold or not.[10] If not, we can rest assured that our original Hookean predictions for the suspicious region remain valid. But if not, we must move to...

Step 3: and insert an altered set of modeling parameters into the region of predicted damage within our original ΔL_H modeling. This is accomplished by inserting three or four additional constants (besides the original E and μ) into our Navier equation modeling that reflect the fact that the rock is no longer isotropic within that little region due to its metamorphosized realignments. To estimate the new parameters required, we must *homogenize* (= average) our ΔL_L conclusions so

---

[9] The specific scale length selected for these submodeling purposes is generally dubbed the RVE (= Representative Volume Element) for the dissection utilized.

[10] As noted previously, I've ignored the thermal degradations between length scales that play a significant role in any realistic modeling of rock damage. We'll discuss such concerns briefly in Chapter 6, but they demand considerations that will otherwise carry us too far afield (Podio-Guidugli 2019).

that the results appear as simple elastic parameters within ΔL_H's smoothed-over setting. In the case at hand, this adjustment only requires that the stress/strain matrix relationships within Navier's equations include a few more numbers than they did previously. In our stage 3 diagram, I have marked these corrective homogenization messages as black feedback arrows demanding alterations in the modeling parameters of our highest level Navier equation modeling. These "homogenization arrows" comprise crucial ingredients within most multiscalar schemes, and I shall expand upon their methodological significance shortly.

Step 4: Because of the local weakening just introduced, our granite block is likely to alter its overall response to the casino's load because of this damaged sector, and so we must completely rework our original ΔL_H calculations once again to obtain a revised estimate $\mathbb{S}_1$ of how the rock will respond to the casino's load. As we noted in the clothesline adjustments of the previous section, the resulting changes may put some fresh regions in danger of metamorphized damage, so we must apply our three-step routine once again to $\mathbb{S}_1$, thereby leading to the estimate $\mathbb{S}_2$.

Thus we repetitiously carry on, continually cycling through the flow chart diagram illustrated until the adjustments required become so small (and near to a hypothesized "fixed point" $\mathbb{S}_\infty$) that we can comfortably stop.[11] Often the final $\mathbb{S}_\infty$ prediction we reach in this manner will differ significantly from our $\mathbb{S}_0$ starting point, perhaps warning us not to build such a heavy building in that locale (from a naïve $\mathbb{S}_0$ perspective, our casino plan may appear

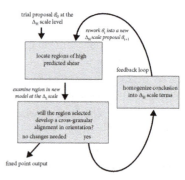

hunky-dory, but our accumulated collection of ΔL_L corrections warns of disaster). As previously noted, these results may require a large number of cycles to produce what is wanted. Nonetheless they usually require immeasurably less time that any straightforward, single-level method of modeling can. Indeed, one can't walk the halls of a contemporary science building without frequently encountering posters

---

[11] These flow chart patterns embody the structuring that I generally have in mind when I write of the "architecture" pertinent to a particular multiscalar reasoning scheme. We should also note that determining when and if these fixed points properly exist can prove a tremendously difficult mathematical task, often addressed in practice by simply applying the program repeatedly to known test cases. Most practical homogenization instructions further incorporate probability estimates of some kind, because we typically lack sufficient information with respect to the original alignments of our microscopic grains to compute an exact elastic constant on the ΔL_H level. These considerations are often quite important in practice, but I will not consider them further, lest they unduly impede the central narrative arc of this monograph.

announcing "A Multiscalar Approach to ...," where the ranges of these ... topics are incredibly varied.

The tremendous efficiencies of multiscalar schemes reside primarily in the fact that each computational stage relies upon relatively simple "dominant behavior" rules with respect to how the material normally responds to altered conditions at the characteristic scale ΔL_H selected. The chore of attending to potential irregularities is delegated to submodels that focus upon the scale lengths ΔL_L where the troublesome damage is likely to first appear (e.g., inside the component mineral crystals that comprise our granite rock). Generally these ΔL_L-level irregularities are driven by local environmental factors that can be captured within our ΔL_L-level submodel as a small-scale boundary or forcing condition that otherwise ignores

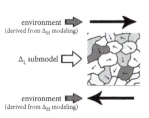

environment (derived from Δ_H modeling)

Δ_L submodel

environment (derived from Δ_H modeling)

what happens in the rock far away. But the technique derives its environmental conditions from the modeling above (viz., ΔL_H) by presuming that its assessment of the ΔL_L modeling element is correct. In the illustration, the two black arrows of applied shear capture this locally induced boundary loading. This factoring of our descriptive task into "induced environment" and "interior response" components is often very effective computationally.

By proceeding in this top-down manner, we manifestly do not attempt to construct "the full story" of everything that happens within our rock, for we never closely scrutinize the behavior of the crystal grain within regions of low stress. Why do so? Nothing important is likely to be found there. Our approach can still prove highly satisfactory in the sense that the disruptive factors that can potentially induce predictive error have been taken into account. But we shall return to this issue of "descriptive completeness" later in the book.

We might also note that the two submodels utilized in this scheme are both equilibrium-focused. We are only worried about the settled "statics" of the rock after it has eventually accommodated the weight of the new casino. If transient factors become important (e.g., earthquake shock waves traveling through the interior), then dynamic submodelings of an evolutionary class might be needed. Even in this situation, passing time might be directly tracked only in the ΔL_H submodel, under the presumption that a finalized lower-scale response will reached quickly enough that the dangers posed by ΔL_L-scale damage can still be monitored using examined an equilibrium-based submodel. As already hinted in the previous section, recognizing these differences in submodel performance will prove critical in our Chapter 6 diagnosis of "cause's" applicational disharmonies.

It is probably obvious that these modeling tactics can be extended to further length scales. We might further worry that high stresses might induce adjustments in interfacial attachment of the sort surveyed in section (ii). Or the material may

become brittle as shifting boundaries gobble up dis-
locations on an even lower-length scale (Murakami
2012). In fact, successful multiscalar architectures
frequently employ a larger number of interacting
submodels. I've added an additional layer of submodel to our granite architecture to illustrate this
possibility.

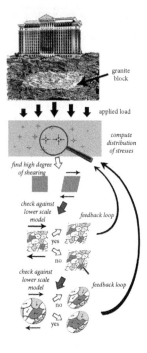

Finally we might observe that important features
of the natural world are registered only within the
flow chart structuring of a wisely conceived multiscalar architecture rather than by overtly appearing
within any of the localized claims computed within
the various submodels. We've already recognized a
chief reason for this absence. When a behavior is
approached in an equilibrium-focused way, we rarely
worry about the frictional factors that allow the target system to shed kinetic energy either in the form of
radiation or internal heating. And we frequently lack
concrete knowledge of how these tribological losses
transpire; we are merely assured with high moral
certainty that they will. This datum alone suffices
for the construction of a highly reliable multiscalar modeling. But the only
acknowledgment of frictional process that appears in the resulting scheme is
located in the manner in which its feedback arrows interconnect its sundry
submodels. That is why I claimed that substantial physical factors sometimes
make their presence known only within in the encompassing architecture that ties
the reasoning scheme together. Typically, a modeling architecture will function
ably only if its arrangements of submodels closely parallel the natural divisions
in "dominant behaviors" that Tao has highlighted and which I have characterized as "natural descriptive opportunities."[12] In Chapter 7, we shall discuss some
of the important ramifications that this otherwise innocuous observation holds
for so-called "scientific realism."

---

[12] I regard these as computational analogs to the opportunities (or "affordances") for environmental
exploitation that nature offers to biological systems. Incidentally, the phenomenon of robust physical
features that are registered only within the architecture of a computational tactic is also common
outside of the range of multiscalar tactics per se. For example, the decision to decompose a piano tone
into standing wave Fourier components rather than traveling wave modes reflects the otherwise
unacknowledged frictional factors that determine how the excited string gradually loses its kinetic
energy (Wilson 2017, 126).

(v)

The secret key to a better resolution of our "Mystery of Physics 101" problems can be found within the *homogenization messages* that shuttle corrective instructions from one submodel to another. Each of the platforms displayed in our sample architecture employs a somewhat different brand of classical mechanics reasoning to reach its local conclusions. For example, ΔL_H treatment presumes that our rock is everywhere smooth and utilizes standard continuum mechanics to reach its estimates of how stresses develop in response to the casino's added weight. ΔL_M divides its consigned segments of granite into anisotropic sectors that can reorient their internal structures in response to local shearing pressures. Finally, our lowest level of submodeling equips its assigned sectors with moveable grain interfaces that can reconfigure themselves as environmental pressures require. Such behaviors demand a more sophisticated form of generalized continuum mechanics to govern these behaviors. The ΔL_H description normally attributes a simple isotropy to a rock locations $\underline{p}$ (elastic constants E and μ) unless instructed otherwise by ΔL_M. But the latter always attributes a more complex set of elastic behaviors (five independent constants) to every location within its jurisdiction, because everything that ΔL_M sees is inside an anisotropic grain. Quite commonly, ΔL_H will continue to attribute unaltered E and μ values to mildly stressed locations $\underline{p}$ whereas ΔL_M demands that more highly stressed regions should be assigned significantly altered material parameters (perhaps $\underline{p}$ falls inside a tourmaline crystal). And neither ΔL_H nor ΔL_M will have any conception of the "configurational forces" that ΔL_L babbles about.

If the descriptive data that each submodel offers are rudely amalgamated together, disagreeable syntactic contradictions immediately arise because the three levels utilize more or less the same vocabularies in locally adapted ways (the four submodels appearing in the billiard ball architecture supplied in Chapter 6 correspond perfectly to the four landings within our Escherian staircase). Intuitively, we probably feel that these divergencies are merely apparent, for each mode of expression "seems right" within its own context. And this is a genuinely appropriate response to which we'll return in Chapter 9. Yes, but try to program a computer to utilize data only when "it seems right"![13] So we must instead institute formal ΔL-associated barriers to prevent the computational conclusions of one submodel from leaking into the remits of the others. But total informational quarantine isn't viable either; our various submodels need to benefit from the wisdoms of their comrades in localized reasoning.

And here resides the simple genius of our "homogenization messages." They represent reformulations of a submodel's ΔL-based conclusions into a linguistic format from which a submodel centered upon ΔL^* can profit without collapsing

---

[13] Such attempts frequently reveal the hidden architectural controls upon which we tacitly rely within our everyday modes of reasoning.

into syntactic contradiction. Our smooth continuum physics level ΔL_H can't directly absorb the anisotropic considerations with which ΔL_M directly deals, but it can understand the direct order "Enter the following five values into your stress/strain matrix at p."

And this where some appeal to "averaging" usually enters the picture. ΔL_M needs to couch its orders in terms of the ΔL_H level "dominant behaviors" in which ΔL_H directly traffics. Level ΔL_M must decide between these "yes" or "no" alternatives: "ΔL_H 'dominant behavior' presumptions look okay with respect to the location p" or "Those claims are out of step with what I can detect, so ΔL_H must be sent a corrective message to alter its elastic parameters." To perform this evaluative function correctly, the homogenization messages should intuitively reflect what the events occurring on level ΔL_M will "look like" when viewed from the perspective of ΔL_H. So we need to erase some of the ΔL_M complexity before any messages are sent onto L_H. But we must do this in a manner that a computer can understand. Unfortunately, mathematics itself is ill-equipped to distinguish a "dominant behavior" from one that is comparatively less important. "What do you mean by 'comparatively less important'?" the subject complains in that robotic voice familiar from old science fiction movies, "The answer does not compute." The most popular device for tricking mathematics into providing the answers we seek is to appeal to "limits." We inflate our original problem into one that involves an infinite population and ask, "What now occurs?" The standard exemplar of this technique (is illustrated by the so-called "central limit theorem" of statistics that extracts a simple bell-shaped curve for the distribution of gains achieved from an infinite population of gamblers.[14] No finite gaggle of wagerers will display the same descriptive simplicity as the infinite ensemble, but the two parameters that characterize the bell-shaped curve (mean and variance) usually supply all of the crucial information that a casino tycoon needs to extract reliable gains from her *sucker sapiens* patrons.

Allied considerations suggest that an appropriate form of "limit" should likewise determine what an ensemble of mineral grains should "look like" upon the higher-scale level of smoothed-over granite appearance. At a most basic level, this account supplies a reasonable picture of how our "homogenization messages" operate. However, further reflection shows that the fuller accounting needs to be more complex. In particular, the "limits" we take must be smart enough to detect when our ΔL_M evaluation needs to send a "revise your parameters" message to level ΔL_M. And familiar forms of "limit" cannot perform this chore correctly. As a result, the technical literature devoted to "homogenization messaging" is often fiercely forbidding. To avoid getting bogged down in refinements, I will exile further discussion of these issues to a later appendix, with some reluctance because

---

[14] This is the exact example that Terrance Tao references in claiming that "the aggregate properties of the system mysteriously become predictable again" (Tao 2012, 24).

of the interesting methodological considerations that they exemplify. At present I will limit myself to a single passing comment.

representative volume element (many randomly oriented grains)

homogenize by shrinking grain size to 0 while keeping the energetic contributions of interior and interfacial gluings constant

Philosophers often write as if the notion of "limit" represents a totally rigid, a priori operation, but this is not true. The "limit"-like construction that proves appropriate for a given application must mimic the actual dominant/subordinate relationships found in nature in a suitable way. Such considerations underscore a lesson that we have already visited: a significant range of physical considerations are only reflected only within the encasing architectonics of a computational strategy and are not explicitly registered within the scheme's component assertions. As a result, the selection of an appropriate "limit" for homogenization frequently represents an insightful art, rather than a thoroughly staid science (Batterman 2021, 90–5).

(vi)

These architectural considerations suggest a better resolution to our "Mystery of Physics 101": stratify our descriptive policies into distinct platforms that share data only through proper homogenizations. We can harness the powers of a formalism (smooth continuum mechanics) that ignores the lower-scale complexities of laminate assembly if we both operate at a scale length where "the aggregate properties of the system mysteriously become predictable again" (Tao 2012, 24) and verify that these assumptions of "dominant behavior" isolation maintain themselves under lower-scale verification. It represents a policy for having one's cake and eating it too, with respect to computational endeavor. As a methodological observation, this claim will seem practically a truism. Thomson and Tait provide a lovely Victorian articulation of the prudent wisdom inherent in allowing lower-scale complexities to be "left out of consideration":

> Until we know thoroughly the nature of matter and the forces which produce its motions, it will be utterly impossible to submit to mathematical reasoning the exact conditions of any physical question. It has been long understood, however, that approximate solutions of problems in the ordinary branches of Natural Philosophy may be obtained by a species of abstraction, or rather limitation of the data, such as enables us easily to solve the modified form of the question, while we are well assured that the circumstances (so modified) affect the result only in a superficial manner.

Take, for instance, the very simple case of a crowbar employed to move a heavy mass. The accurate mathematical investigation of the action would involve the simultaneous treatment of the motions of every part of bar, fulcrum, and mass raised; but our ignorance of the nature of matter and molecular forces, precludes any such complete treatment of the problem.

It is a result of observation that the particles of the bar, fulcrum, and mass, separately, retain throughout the process nearly the same relative positions. Hence the idea of solving, instead of the complete but infinitely transcendent problem, another, in reality quite different, but which, while amply simple, obviously leads to practically the same results so far as concerns the equilibrium and motions of the bodies as a whole.

The new form is given at once by the experimental result of the trial. Imagine the masses involved to be perfectly rigid, that is, incapable of changing form or dimensions. Then the infinite series of forces, really acting, may be left out of consideration; so that the mathematical investigation deals with a finite (and generally small) number of forces instead of a practically infinite number. Our warrant for such a substitution is to be established thus.

The effects of the intermolecular forces could be exhibited only in alterations of the form or volume of the masses involved. But as these (practically) remain almost unchanged, the forces which produce, or tend to produce, them may be left out of consideration. Thus we are enabled to investigate the action of machinery supposed to consist of separate portions whose form and dimensions are unalterable.

Enough, however, has been said to show, first, our utter ignorance as to the true and complete solution of any physical question by the only perfect method, that of the consideration of the circumstances which affect the motion of every portion, separately, of each body concerned; and, second, the practically sufficient manner in which practical questions may be attacked by limiting their generality, the *limitations introduced being themselves deduced from experience*, and being therefore Nature's own solution (to a less or greater degree of accuracy) of the infinite additional number of equations by which we should otherwise have been encumbered.   (Thomson and Tait 1912, I 1–3)[15]

The only surprise lies in the fact that tremendous predictive advance can be achieved if these basic policies of higher-scale predication tempered by lower-scale correction are automated and run through repeated loops until the results stabilize upon fixed values. Each level of submodel is allowed to "do its own thing" with respect to its parochial domain of behavior rules, trusting that any departures from these presumptive regularities will be noted in submodels devoted to the

---

[15] As previously noted, Thomson and Tait engage in a number of misty claims to the effect that "it's all in Newton" that are hard to reconcile with the present observations.

scale levels where the disturbing factors first arise (an unusual macroscopic decline in elastic capacity is presaged by boundary reattachments that are dominant at the mesoscopic grain level).

In the previous chapter, we reviewed Hertz's developmental account of how hidden conceptual tensions come to inhabit a descriptive process. Borrowing directly from Hertz,[16] David Hilbert described these processes as follows in his celebrated "6th problem on mechanics":

> The physicist, as his theories develop, often finds himself forced by the results of his experiments to make new hypotheses, while he depends, with respect to the compatibility of the new hypotheses with the old axioms, solely upon these experiments or upon a certain physical intuition, a practice which in the rigorously logical building up of a theory is not admissible.   (Hilbert 1902, 250)

Like Hertz, Hilbert presumes that the proper corrective for the resulting tensions lies in uniform axiomatization, a resolution we are now disputing. If restricted communication pathways can be established between our sundry submodel "protectorates,"[17] then it is less clear that anything truly "inadmissible" appears in our suitably tempered procedures. Indeed, finding oneself driven to the utterly implausible hypotheticals required within Hertz's single-leveled axiomatics strikes me as highly undesirable, if not "inadmissible." As noted before, soon after Hilbert articulated his 6th problem, physicists began to attend primarily to quantum theory, with its peculiar dependence on the simple ODE operators encountered within a Euler's recipe construction. It is largely this shift in attention that obscured the brute realization that an axiomatic regularization of the full dimensions of real-life classical modeling practice is apt to prove quite weird.

Here's an analogy that illuminates the developmental considerations that Hertz and Hilbert have in mind. A descriptive practice gradually consolidates into localized grains from an amorphous original melt. As this happens, the enlarging grains jam against one another, forming laminate boundaries. In this way several localized protectorates emerge, generating discordances along their bordering edges. Within each localized patch, a word such as "force" will specialize to accommodate parochial demands, rather as new species emerge when scattered populations become segregated into isolated tribes and adapt to local environmental advantage

[16] Hilbert is implicitly characterizing Hertz's point of view, which came first and partially inspired Hilbert's own interests in axiomatization.
[17] I learned this evocative phraseology from Bob Batterman.

(Mayr 1997). Wilson (2006) supplies a large repertory of patchwork usages of this character.

Our everyday use of "force" often exhibits patch-centered specializations of this kind. For example, upon a macroscopic scale engineers characterize the retarding effects of adjacent layers within a viscous fluid as due to a shearing force V across the sheets.[18] When we drop to a microscopic scale, some of this erstwhile "force" becomes instead allocated to molecular transport, adjusting "force's" local attachment to just the action-at-a-distance interactions between the component units. Such reattachments spontaneously aspire in practice, and their applicational disharmonies are scarcely noticed. This policy creates no problems unless someone attempts to naïvely drag the upper-scale V into molecular confines.

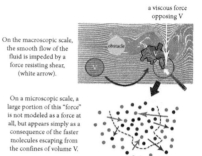

a viscous force opposing V

On the macroscopic scale, the smooth flow of the fluid is impeded by a force resisting shear, (white arrow).

On a microscopic scale, a large portion of this "force" is not modeled as a force at all, but appears simply as a consequence of the faster molecules escaping from the confines of volume V.

In Wilson (2006), I ascertained these patchwork boundaries largely by considering the *classificatory employments* of the terms in question. Inspired by the achievements of modern computing and by attending to the *reasoning pathways* along which information is shuttled from one region to another, I realized these homogenized interconnections frequently have the net effect of stacking one patch above another in the characteristic manner of a multiscalar architecture. This simple layering via feedback messaging converts an adjusting word usage that threatens to become a harmful ambiguity into a powerful weapon for combating Tao's "curse of dimensionality."

In fact, these are essentially that same safeguards that allow Hertz's "forces which arise through motion" and "forces which are a consequence of motion" to work together in descriptive harmony. A closer inspection of usage reveals an appeal to what we might call "asymptotically separated scale lengths." In a standard approach to our beads on a wire problem, the beads are treated as "material points" while the constraining wire is captured as a rigid curve (hence the asymptotics: small bead → 0-dimensional "point"; larger wire → 1-dimensional line).[19] In the previous chapter I provided the following diagram to illustrate this implicit separation in scales. Note that the action-at-a-distance "forces" between the bead-points are regulated upon the lower N_L scale using Euler's recipe rules (rendering them "forces which arise through motion"). In contrast, their

---

[18] See Wilson (2017, 26–7). Note that these considerations are closely allied to the problems of mass conservation discussed in our appendix on "Historical Complexities."

[19] These reductions in dimension play a purification role closely related to the infinite populations utilized in our homogenization policies.

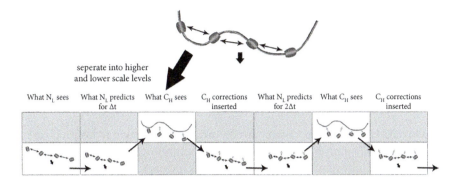

upper-scale (C_H) constraint "force" corrections are computed geometrically ("as a consequence of motion") based upon the beads' displacements with respect to the wire. Observe that this higher C_H information (wherever they wander, the beads will stay on the wire)[20] represents *partial information* that we can effectively exploit as a means of avoiding the unknown (and horribly complex if feasible at all) collection of Newtonian forces that would otherwise be needed to keep the beads within a close proximity of the wire. Why should we re-derive what we already know? Textbooks utilize so-called "Lagrange multipliers" to shuttle information between these scales in the general manner of a homogenization message (Wilson 2017, 320–3). The zig-zag interconnections in our diagram between N_L and C_H levels correspond to the "successive approximation" computations usually needed to locate the "fixed point" stabilizations of arrangement like this (*vide* our earlier approach to a sagging clothesline). In the next chapter, we'll see that such computational policies supply a vital symptom of an important subclass of explanatory patterns.

Multiscalar architectures reveal the secret methodology concealed within our everyday descriptive policies. My college physics teacher had attempted to initiate me into the navigational quirks of this sprawling landscape but neglected to explicate the cartographic reasons for its unexpected twists and turns. In effect, I needed to walk upon a descriptive landscape fabricated in the manner of an Escher's staircase, but neither my instructor nor my brother's philosophy of science books warned me of that fact. But I later discovered that I was in good company. Heinrich Hertz was coaxed into axiomatic impracticality by a similar presumption that the same business ought to become neatly mounted upon a flat and self-consistent plane.

In miniature, our diagnosis of "force's" errant behaviors typifies a general approach that can aptly be applied to a considerable range of philosophical puzzles. We must sniff out the underlying practicalities that secretly contort a

---

[20] I echo John Wayne in the film *Red River*: "We don't need no fences because wherever they go, they'll be on my land."

puzzling form of discourse into complicated sectors. It represents a diagnostic model we shall apply in several additional circumstances in our remaining chapters.

(vii)

Although Hertz did not envision our non-axiomatic resolution to his problem of doctrinal overloading, his views of the developmental origins of such conflicts are very much in line with what I advocate. Let's look once again at the crucial passage:

> But the answer which we want is not really an answer to the question [of the nature of force]. It is not by finding out more and fresh relations and connections that it can be answered; but by removing the contradictions existing between those already known, and thus perhaps by reducing their number. When these painful contradictions are removed, the question as to the nature of force will not have been answered; but our minds, no longer vexed, will cease to ask illegitimate questions.   (Hertz 1894, 7)

As noted in Chapter 3, the final sentence strongly spurns individual conceptual analysis in Bertrand Russell's "classical" manner as untrue to the fashion in which we have learned our words. We do not dwell intently on the inner meaning of "force" and apply it subsequently to the world, engaging in flubs only when we forget the original conception with which we started. Instead, we begin with a rough plunge and gradually refine our strokes as we progressively swim about in the grand swimming pool of practical application. Without noticing that we have done so, we may tacitly internalize the realization that breaststroke is best when we find ourselves in sector A; the Australian crawl in sector B. These are applicational skills that we can verifiably learn, whereas Russell's form of intensive "conceptual grasp" constitutes a semantic pipedream, a version of Quine's "myth of the mental museum."

In many respects, Hertz himself might have found our alternative reconciliation of doctrines attractive, once he realized that the resolution of his concerns with respect to transverse waves lie in pure electromagnetism beyond the reach of traditional classical doctrine. After all, he looks upon mechanical practice as a functioning whole with a few misaligned parts. Our multiscalar programmers have merely discovered a superior kind of machine that Hertz had overlooked.

This holism conception of the descriptive enterprise stems from a desire that Hertz shares with many of his philosopher/physicist contemporaries, notably including Ernst Mach. Productive science needs to be liberated from the overbearing demands of traditional "big metaphysics" (which merely systematizes the inherited tactics of everyday "instinctive knowledge"). According to Mach (and presumably Hertz), claims that strike us as "intuitively true" merely reflect psychological rules of thumb we have inherited from past generations and which have been vigorously reiterated within our early linguistic training. Mach acknowledges that these reasoning directives often remain valuable in initiating our advancement along the winding and improving byways of useful language, but they nonetheless remain of limited reliability applied beyond those initiating prods:

> While, now, I wish to emphasize on the one hand that every idea which can help, and does actually help, is admissible as a means of research, yet on the other hand it must be pointed out how necessary it is from time to time to purify the exposition of the results of research from the unessential ingredients which have become mixed with them by working with hypotheses. For analogy is not identity; and for complete understanding we must have, besides knowledge of the similarities and the agreements, knowledge of the differences as well. When I try to do away with all metaphysical elements in the exposition of natural science, it is not my opinion that all ideas which are meant to serve as images are to be put aside, if they are useful and are viewed merely as images. Still less is an anti-metaphysical critique to be regarded as directed against all fundamental principles which have hitherto show themselves to be valuable. We may, for example, quite well have strong objections to the metaphysical conception of "matter", and yet not necessarily have to eliminate the valuable conception "mass": we can retain this latter conception as I have done in my Mechanics, because we have seen that it signifies nothing more than the fulfillment of an important equation.   (Mach 1896, 333–4)[21]

As it happens, Hertz is somewhat less wary than Mach with respect to the incursions of "metaphysical dogma," as we'll observe in Chapter 9. I happen to share Hertz's tolerance with respect to the realms of "small metaphysical concern," considerations that we'll review in the same locale.

The success of our multiscalar resolution opens a door to a gentler approach to the ongoing tasks of conceptual diagnosis, one that is better equipped to deal with conceptional disharmonies as they arise *media vita*. After a careful study of "beads on a wire" circumstances, we can figure out how their asymptotically separated

---

[21] See also Helmholtz:

[T]his is however in disagreement with the older concept of intuition, which only acknowledges something to be given through intuition if its representation enters consciousness at once with the sense impression, and without deliberation and effort.... I believe the resolution of the concept of intuition into the elementary processes of thought as the most essential advance in the recent period.   (Helmholtz 1878, 130 and 143)

descriptive scales can be persuaded to systematically work together with the assistance of some Lagrange multiplier homogenizations. We needn't simultaneously resolve every conceptual clash encountered across the wide landscape of classical descriptive policy in the manner of Hertz's once-and-for-all axiomatic resolution. Indeed, we have very good reasons for doubting that a unified repair of this sort exists. When we follow the typical pathways of "lousy encyclopedia" deferral, we eventually find ourselves dipping into articles that reek strongly of quantum mechanics. If we excise all of these non-classical steppingstones, we are likely to find ourselves stranded upon a peculiar archipelago arranged like an Escherian staircase, for which no single-leveled axiomatization is plausible (Wilson 2006, 195–7).

If we abandon these utopian ambitions, we find that we can still crack refractory conceptual nuts with a fair degree of success and finesse. Indeed, that is what I think we have just achieved with respect to the venerable "Mystery of Physics 101." Hertz represents the observant gumshoe who first spots the essential clues, and computer scientists serve as the Perry Masons who eventually piece together the criminal enterprise.

(viii)

To be sure, in tolerating multiple platforms within classical mechanics in this adjustable manner, we deprive ourselves of our ability to address our Chapter 2 question of "whether a transverse wave ether is mechanically feasible?" with the same degree of full "inductive warrant" definitiveness that Hertz hoped to obtain. As we noted, different choices of Escher staircase "foundations" can supply divergent answers on this score. But is this happenstance so dreadful? As Poincaré noted, we find ourselves in a universe that contains neither a ponderous ether nor fully operates by classical principles of any kind. But as that happens, we also lose any "inductive warrant" from nature that some satisfactory single-leveled axiomatization of standard classical practice will prove feasible. Carnap's assurances to the contrary, presuming that these fictitious completions actually exist obscures the significant roles that the imperfect safeguards of multiscalar organization play in restraining our engines of improving reasoning technique from running off their rails. Carnapian obfuscation will only hinder our diagnostic endeavors, for it is exactly within these same "imperfect safeguards" that the characteristic symptoms of conceptual discord make themselves manifest.

Let me once again stress that less ambitious forms of axiomatic specification are indubitably useful in more guarded investigations of reasoning architecture. The fact that purely transverse waves cannot be readily implemented within smooth continuum mechanics is good to know. It is likewise illuminating to learn that a given phenomenon violates "material frame indifference" (Truesdell 1977), even when we know of acceptable "classical behaviors" that violate this condition. But many of the assertions that philosophers commonly advance with respect to an otherwise unspecified "classical mechanics" cannot be coherently defended without appeal to a satisfactory form of all-encompassing axiomatization which simply does not appear to exist. We will discuss these issues further in Chapter 8.

Overwrought or premature emphases upon axiomatization possess a venerable history that begins long before the episodes I have cited from the 1930s. In Chapter 3, we cited Pierre Duhem's jingoistic disdain for "the ample but weak mind of the English physicist." Duhem's diatribe praises the high-minded axiomatic rigor of Continental physicists such as himself:

The theories created by the great Continental mathematicians, whether French or German, Dutch or Swiss, ... offer a common feature: they are understood to be systems constructed according to the rules of strict logic. Products of a reason unafraid of profound abstractions or long deductions, but mainly eager for order and clarity, their theories demand that an impeccable method characterize the series of their propositions from beginning to end, from the basic hypotheses to the consequences that can be compared with the facts. It was this method that brought forth those majestic systems of nature claiming to bestow on physics the formal perfection of Euclid's geometry. These systems take as their foundations a certain number of very clear postulates and try to erect a perfectly rigid and logical structure in which each experimental law is exactly lodged.... [S]uch edifice[s have] stood as the perpetual ideal of abstract intellects, especially of the French genius. In pursuing this ideal it has raised monuments whose simple lines and grand proportions are still an object of delight and admiration, especially today when these structures are shaky on account of their generally undermined foundations. This unity of theory and this logical linkage among all the parts of a theory are such natural and necessary consequences of the idea that strength of mind imputes to a physical theory, that to disturb this unity or to break this linkage is to violate the principles of logic or to commit an absurdity, from its viewpoint.

Not at all like that is the case of the ample but weak mind of the English physicist. Theory is for him neither an explanation nor a rational classification of physical laws, but a model of these laws, a model not built for the satisfying of reason but for the pleasure of the imagination. Hence, it escapes the domination of logic. It is the English physicist's pleasure to construct one model to represent one group of laws, and another quite different model to represent another group of laws, notwithstanding the fact that certain laws might be common to the two groups.... Thus, in English theories we find those disparities, those incoherencies, those contradictions which we are driven to judge severely because we seek a rational system where the author has sought to give us only a work of imagination.   (Duhem 1914, 80–1)

The purist emphases that characterize Clifford Truesdell's otherwise excellent writings on "rational mechanics" represent an allied prolongation of these "rigorist" proclivities (no longer accepted wholeheartedly by researchers in this field—see the appendix to Chapter 3).

A more sympathetic reading of the long quotation from Thomson (= Lord Kelvin) and Tait in section (vi) reveals a sensitive concern with maintaining the ongoing reliability of science in a manner that is frequently absent in Duhem-like panegyrics to theory T rigor. By deftly exploiting a higher-scale constraint or appealing to an observational generalized coordinate the pitfalls of excessive speculation can often be skirted. Granted, the "disparities, incoherencies and contradictions" that emerge when they attempt to deal with the rock-bottom "foundational issues" of their science are very frustrating, for they sometimes pretend to "rigor" when nothing of the sort is evident.[22] But their palpable awareness of the methodological checks and balances to be cultivated in developing an increasingly reliable body of science should be recognized as something other than a peculiar nationalist quirk.

APPENDIX

Further Comments on Homogenization

In the scientific literature devoted to multiscalar architecture, a suitable choice for "homogenization messaging" is often fiercely forbidding, for technical considerations that I shall sketch in a moment. Nonetheless, a suitable format of messaging between scale lengths ΔL and ΔL^* usually comprises an intuitively plausible answer to the question, "What do the events occurring at a scale level ΔL^* *look like* when viewed from the characteristic length scale ΔL?" And we enjoy a preliminary sense of what these answers should look like: upon the macroscopic ΔL_H level of the granite block, its ΔL_L-level grain adjustments will look like a patch of anisotropic damage within the rock, which otherwise appears as an unvariegated smooth continuum (ΔL_H can't "see" any of ΔL_L's component grain). As such, this talk of "perspective relative to a characteristic scale length ΔL" metaphorically recapitulates Terrance Tao's observation that upon certain structural levels "the aggregate properties of the system mysteriously become predictable again." In Chapter 8, we shall consider another interesting variant within this general "ordinary language manner of talking about homogenization relationships" class.

However, the question of exactly how these limits should be taken is a quite delicate affair that needs to parallel the physical emergence of dominant/subordinate behaviors in a natural way. The threshold behaviors we discussed with respect to laminate interfaces within granite illustrate a typical problem. As we saw, as the environment pressures

---

[22] These foibles are evident in their steadfast loyalty to Newton's original words (presumably because, in spite of all temptations to belong to other nations, they remain Englishmen) (*HMS Pinafore*).

about a small chuck of rock become severe, the specimen will find it energetically advantageous to rework the patterns in which the component grains are cemented together instead of resisting the impressed shear entirely through elastic strain within the crystals themselves. But reforming the interfacial bonds between the grains demands a certain degree of chemical activation energy which creates a threshold effect in which the boundary adjustments only appear after the environmental stresses have reached a critical level. These same considerations determine the appropriate "yes" or "no" messages that our laminate level ΔL_L examination should pass along to the smooth continuum submodeling employed at the higher ΔL_H scale (*yes*: retain your original Hookean constants E and μ below the threshold T; *no*: replace those values with revised anisotropic factors above T). Unfortunately, conventional forms of averaging completely ignore whatever happens on structures that are of lower dimensionality than the volume over which one is "averaging" (i.e., integrating). But this is exactly the situation we confront with respect to a laminated material; the grain interfaces are of dimension two (often symbolized as ∂V) while the mineral interiors they surround (V) are three-dimensional in character. Untutored mathematics will ignore the energetic thresholds upon these ∂V grain borders unless it is supplied with specific instructions to attend to these regions in a better way. Doing so requires that we fuss with our initial notions of "integration" and "limit" in rather fancy ways, a task which typically involve puffing up the ∂V regions until they assume a status more coequal with their interiors.

Allied tactics are commonly required when we consider the familiar policy of dimension reduction, as when we attempt to model a symmetrical prismatic volume as a two- or even one-dimensional elastic "beam." Classical paradoxes of mismatch commonly arise when the "limits" we naturally apply to boundary and interior regions wind up not agreeing with one another (the reduced boundary limit places requirements upon the reduced interior that the latter cannot satisfy coherently). A careful reexamination of these intuitive "limits" using the tools of functional analysis reveals the presence of a previously unanticipated "boundary layer" in which the model's boundaries and interiors need to find their accord.

In fact, refined homogenization studies sometimes replace the blunt "yes or no" evaluations of upper-scale behavior that we have surveyed with subtler forms of "intermediate asymptotics" linkage employing so-called "handshake methods" (Batterman 2021). The computational interconnections that we find ourselves able to fashion in this way do not represent subjective whimsies on our part but directly reflect the hierarchies of descriptive opportunity that nature makes available to us when we possess the wit to recognize them. A well-devised multiscalar architecture mimics these external arrangements within its internal policies of homogenized message exchange.

Such refinements return us to the "intuitive guesses" that usually play an important developmental role in the articulation of a viable multiscalar architecture. The traditional disparities between boundary and interior descriptions within traditional beam theory were originally reconciled in this way (e.g., by appeal to "Saint-Venant's Principle," later followed by more detailed examples involving series truncation and asymptotics, even later followed by rigorous homogenization treatments employing subtle forms of functional analysis) (Kennedy and Martins 2012). The latter can be profitably carried out only after the intuitive relationships have been mapped out through a fair amount of trial-and-error experimentation.

> For it is in mathematics just as in the real world; you must observe and experiment to find the go of it.... All experimentation is deductive work in a sense, only it is done by trial and error, followed by new deductions and changes of direction to fit

circumstances. Only afterwards, when the go of it is known, is any formal explication possible.   (Heaviside 1912, II 33)

I have sometimes dubbed the unavoidable imperatives that require that topic A be roughly mapped out before any investigation of topic B can profitably commence as the "season-alities" of investigative endeavor. Sometimes we need to accurately guess before we're remotely ready to rigorously homogenize. After we have done so, we can return to our original guesses and improve them through after-the-fact tinkering.

One of my central complaints with respect to philosophy's current emphasis on "ersatz rigor" is that such thinking conceals all of these tell-tale "seasonalities" beneath the floorboards of a "fictitiously completed theory T." Such accounts fail to acknowledge the gradualist manners in which we gradually learn to convert an amorphous descriptive practice into a more sharply articulated reasoning architecture containing many intercom-municating rooms. Demanding fully transparent interconnections from the get-go repre-sents an inappropriate demand of the kind that when Oliver Heaviside rejects when he spurns "the wet blanket of the [mathematical] rigorists." Indeed, within many real-life circumstances the slogan "science always idealizes" operationally signifies a tentative suggestion that can be more accurately captured as reasonable apologetics in favor of educated guesswork. At the price of elegance, the policy is better captured as: "science often selects a reduced complexity situation that it hopes will eventually emerge as a significant 'dominant behavior' submodel within a properly homogenized encompassing architecture."

6

Diversity in "Cause"

Man is governed by the struggle for self-preservation: his whole
activity is in its service and only achieves, with richer resources,
what the reflexes accomplish in the lower organisms under simpler
conditions of life. Every recollection, every idea, every piece of knowledge has a value originally only in so far as it directly furthers man in
the direction indicated. The life of ideas reflects the actual facts,
supplements partially observed facts according to the principle of
similarity (by association) and makes it easier for man to place himself
in more favorable relations to them. The more extensive the field of
facts and the more truly that field reflected, the more exactly ideas are
adapted to the facts, the more effectively helpful are those ideas in life.
But only what most powerfully concerns the will, the interest (that is
the useful), or what stands out strikingly from the frame of daily life
(the new, the wonderful) will initially attract attention. Only gradually, from this point, are ideas able to adapt themselves to broader
fields of facts. Here the continuous widening of experience, often
resulting from chance circumstances, plays an essential part....
Should anyone be prone to think that the foregoing discussions are
no subject for a scientific public, he is mistaken; for science is never
severed from the life of the every-day world. It is a flowering of the
latter and permeated with its ideas.

Ernst Mach (Mach 1896, 336–7)

(i)

The focal concerns of this monograph are directed towards philosophical methodology generally, not simply to a half-forgotten problem from the nineteenth
century physics. So let us turn to a controversy of greater contemporary salience:
the "metaphysical" status of the notion of "cause." Here we witness a wide range of
divergent opinions, ranging from the contention that any philosophical emphasis
on "causation" merely reflects an undesirable retention within modern science of
primitive views of natural process, to the belief that "causation" comprises a
fundamental metaphysical category underlying the very manner in which

Imitation of Rigor: An Alternative History of Analytic Philosophy. Mark Wilson, Oxford University Press. © Mark Wilson 2022.
DOI: 10.1093/oso/9780192896469.003.0006

scientific inquiry proceeds. We shall examine both of these doctrines in the pages ahead, as well as several other approaches that avoid their stark extremes.

As it happens, many of the philosopher/scientists of the nineteenth-century era endorsed the first thesis strongly because they were concerned to reject the conservative restrictions upon "the conceptual liberty of the scientist" that allegedly followed from some dubious "universal law of causality" endorsed by their "metaphysical" critics. Ernst Mach, for example, maintained that the straightforward empirical investigation of thermal behavior had been seriously compromised by an aprioristic demand that an atomist "causal process" substratum be supplied for this subject. Following Mach, the British physicist J. H. Poynting complained as follows in 1899:

> It would be a very real gain, a great assistance to clear thinking, if we could entirely abolish the word "cause" in physical description, cease to say "why" things happen unless we wish to signify an antecedent purpose, and be content to own that our laws are but expressions of "how" they occur.
>
> The aim of explanation, then, is to reduce the number of laws as far as possible, by showing that laws, at first separated, may be merged in one; to reduce the number of chapters in the book of science by showing that some are truly mere sub-sections of chapters already written.   (Poynting 1899, 385–60)

Authors of this inclination typically favor a simple inductive empiricism of a Humean "constant conjunction" character, in which legitimate talk of "causes" and "effects" merely reflect well-confirmed generalization of the form "All events A occurring at time t are succeeded by effects B occurring at time t + Δt." Variants upon this Humean position remain popular to the present day.

In order to dispel traditionalist claims that talk of "causation" captures some primordial "cement of the universe," Mach and his followers claimed that these intuitive intimations of something metaphysically grander merely represent unhelpful intrusions from ancient animism and/or childhood self-centeredness. Yes, our childhood uses of the word "cause" emerge in contexts laded with quasi-anthropomorphic associations, and these lingering connotations embellish our Humean patterns of talking of "causes" with a residual veneer of misleading aspect. Somewhat uncharacteristically, J. L. Austin subscribes to such a "surviving ancient models" thesis:

> Going back into the history of a word,... we come back pretty commonly to pictures or models of how things happen or are done. These models may be fairly sophisticated and recent,... but one of the commonest and most primitive types of model is one which is apt to baffle us through its very naturalness and simplicity.... A model must be recognized for what it is. "Causing", I suppose, was a notion taken from a man's own experience of doing simple actions, and by primitive man

every event was construed in terms of this model: every event has a cause, that is, every event is an action done by somebody—if not by a man, then by a quasi-man, a spirit. When, later, events which are not actions are realized to be such, we still say that they must be "caused", and the word snares us: we are struggling to ascribe to it a new, unanthropomorphic meaning, yet constantly, in searching for its analysis, we unearth and incorporate the lineaments of the ancient model. As happened even to Hume, and consequently to Kant. Examining such a word historically, we may well find

Austin

that it has been extended to cases that have by now too tenuous a relation to the model case, that it is a source of confusion and superstition.   (Austin 1966, 150–1)

In the opening epigraph to this chapter, Mach gives voice to a similar diagnosis which attributes a further portion of these unwanted supplementations to the fact that young children first master the word in the context of activities that they themselves initiate (thus Mach remarks, "only what most powerfully concerns the will . . . initially attracts attention").

Certainly some degree of justice resides within these remarks. But the criticisms that other parties to this dispute lodge against their rivals have their merits as well, lending the ensuing debate the appearance of a circular firing squad. We have already witnessed similar contrastive patterns in Chapter 5's investigation of "force"; each of its varying localized specializations appear to leave out some conceptual ingredient that plays a significant role in its usages elsewhere. Some employments of "force" appear to demand "force law" antecedents (gravitation and the Newtonian attractions between our beads) and some do not (the reaction "forces" derived from higher-scale constraints). Indeed, Hertz distinguishes these two classes of "force" precisely on the grounds that they invoke notions of "causal responsibility" that operate in different patterns.

In fact, the confusing behaviors of "cause" strike me as similar in their developmental origins to those that attach to "force," which isn't altogether surprising due to the fact that the applicational demands upon the two words are closely linked. To see this, we shall examine a prototypical multiscalar architecture in which several of the popular specializations of "cause" naturally appear, sheltered within their localized pockets of homogenized message protection. Attempting to find some common "semantic" basis uniting them all in simple polysemic variation is as unlikely to produce univocal results as an allied quest for a base meaning for "force" that can submit to straightforward axiomatic regulation. The backgrounded multiscalar architecture that connects together our localized specializations of "cause" likewise contains a hidden Escherian twist that obstructs a single-minded "analysis" of this same character.

In the rest of this chapter, we shall scrutinize our ordinary patterns of talking about "causes" and "effects" with respect to everyday billiard ball behavior from multiscalar perspective. We can then appreciate how the shifting patterns of applicational focus characteristic of these words trace to the underlying adjustments in reasoning strategy that are secretly responsible for the remarkable efficiencies of their usage. In the chapter to follow, we shall draw upon this example as a means of illustrating the unhelpful restrictions that ersatz conceptions of "rigor" impose with respect to developing a more sympathetic unraveling of these puzzling words.

(ii)

Before we commence our case study, let us first reflect upon some of the terminological demands that necessarily accompany any backgrounded reliance upon multiscalar controls, whether we are consciously aware of these structural demands or not. As long we are only concerned with descriptive vocabularies that readily submit to single-leveled axiomatic organization, their supportive "semantics" can sometimes be adequately explicated in the simple patterns of word-to-world alignment that are characterized as "Fido"/Fido arrangements.[1] In such cases, a significant portion of a word's intuitive usage can be adequately captured within these simple forms of direct "referential" support. But these relatively unproblematic vocabularies are not the usages that prompt Hertz's and Mach's deeper insights into the manner in which important words progressively gain firmer attachments with the physical world as their practical capacities enlarge and diversify. In such cases, words like "force" or "cause" will gradually condense from an amorphous primordial melt and specialize into localized pockets ("submodels" in our technical exemplars) that communicate with one another through controlled lines of communication. For these adaptations to operate successfully, additional terminologies of a recipe-formative character are needed to initiate novices into the applicational byways of the emerging patchwork. Students in Physics 101 need to somehow absorb instructions of the form, "In computing the direct forces between the beads on a wire, follow procedure A, but when calculating the constraint forces attributable to the wire, follow procedure B. Then combine your results in pattern C." These are imperatives, rather than straightforward assertions. In navigation, the linguistic instruction "First turn left, then right" does not directly describe the world, but tells you how to find your way through it. Just so, the proper usage of multiscalar descriptive terms sometimes requires procedural instructions of an allied sort. With respect to the buried granite of the previous chapter, we might

---

[1] The term derives from Gilbert Ryle (Ryle 1953, 167), but I reject the anti-referential satire that he intended. However, he is correct in suggesting that such a treatment alone cannot adequately capture the physical determinants of the words we are presently considering.

need to direct a beginner in soil resettlement as follows. "To reason profitably about a large block of granite, first *map out* a trial behavior utilizing simple Hooke's law considerations to gain a rough sense of how applied stresses are likely to distribute themselves across the rock, presuming that no significant anti-Hookean damage arises anywhere. As a crosscheck on the validity of this assumption, *subsequently adjust your modeling focus and examine* whether the predicted stresses are *actually sustainable* in view of the granular structure of the rock upon closer analysis. And similar *inquiries should be prosecuted* to several depths of *length scale*. Don't become confused when your various stress assessments appear to disagree with one another; your results will need to be *properly averaged* before they can be *directly compared* in that way." The italicized phrases capture the procedural recipe that our pupil must learn in order to successfully implement the checks-and-balances scheme we have outlined. Considered solely with respect to these imperative functions, the italicized specifications do not directly reflect the physical features of the granite itself (except indirectly, in the sense that good instructions for making banana bread tell us *something* about bananas themselves). So let us characterize terminologies of this recipe-focused character as "procedural" or "managerial," for their primary function is to enforce the manner in which the submodels in our architecture must be correctly interconnected and revised. Observe that sometimes we must explicitly proscribe the homogenization policies that should be followed in exporting data from one local patch to another. This specific managerial role will emerge as especially salient philosophically in Chapter 8.

Evolving usages frequently exhibit a mixture of managerial and descriptive characteristics when they specialize from a less disciplined parent usage in this manner. "Force" illustrates both characteristics amply. But our term of current interest ("cause") operates in much the same manner. So let's first examine the managerial purposes it often facilitates.

Consider the causal claim we might make to a young child.

You *caused* the cat to scratch when you pulled its hair.

Detailed psychological studies have amply verified the claim, already pronounced in Mach's commentaries, that a child's first employments of the word are strongly fashioned upon an egocentric basis, frequently coupled to some measure of approbation or fault (Bonawitz et al. 2010). These agentive origins have spurred Mach, Austin, and other critics to the supposition that these vestigial connotations are responsible for the illusion that talk of "cause" within science demands richer descriptive underpinnings than are provided within Hume's nakedly empiricist emphasis upon the constant conjunctions of occurrent fact. In this manner, they seek to explain away the intuitive trepidations to the effect that Humean approaches can't render proper justice to our conception of "causation" (this is a contention we will revisit shortly).

Although anthropomorphic strands undoubtedly survive within our adult employments of "cause" as a variety of latent DNA (Wilson 2006, 352), I don't believe that such lingering ingredients play any significant role in generating the familiar "circular firing squad" aspects of the word with which we will find ourselves concerned. I will instead emphasize another aspect of its usage that also emerges relatively early in juvenile learning, viz., its *managerial* role in articulating a *procedural recipe*. In particular, let us examine the manner in which talk of "causes" and "effects" can serve to mark the potential *obstacles* that we might confront in completing a recommended recipe. Suppose a child is learning how to set up a long chain of dominoes in which an initial push will sequentially collapse the entire chain. We might identify a lacuna in the child's setup recipe to a child as follows:

You didn't push the first domino hard enough to make this other domino down the line topple over in a way that will successfully cause the next domino to fall.

In doing so, we erase any original agentive attachments to the word by applying the term to autonomous later events that the child does not directly control (Woodward 2021). In so doing, "cause" shifts to isolating a critical intervening event within the course of an unfolding process. A pupil so instructed will readily categorize a sequence of "causes and effects" that transpire on a pool table in similarly semi-autonomous terms ("semi-autonomous" because some initiating agent still needs to yield the cue stick). I believe that the utilities of highlighting a sequence of events in "causal terms" quickly overwhelms any lingering anthropomorphic ingredients.

Following this same developmental course, "cause" further evolves to serving as a "critical marker within a sequential recipe." Here's part of an instruction for drawing a cartoon duck:

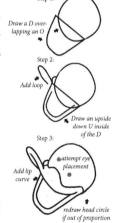

Step 1:
Draw a D overlapping an O

Step 2:
Add loop

Draw an upside down U inside of the D

Step 3:
Add lip curve

attempt eye placement

redraw head circle if out of proportion

Draw an oval and a big D, for the head and beak respectively. However, if you make the second figure too large, it'll cause you to place the eyes in the wrong place, so adjust its dimensions until the two dots for eyes seem in the right place.

Here the term "cause" merely identifies a consideration that arises as an "obstacle" to be confronted as an agent attempts to complete the second stage of the suggested recipe. Plainly no weird psycho-physical interaction between yourself and the drawing paper is implied.

Indeed, the successive approximation recipe utilized in the previous chapter's approach to estimating the shape of a sagging clothesline can be articulated in these "obstacle to be confronted" terms as well:

> To draw (or calculate) the shape of a loaded clothesline, first calculate the sagging **effect** that the heaviest piece of clothing will **cause** to occur. Perhaps this heaviest item is the dress, weighing x lbs. Make sure that it **causes** the line to bend just enough that it can bear the dress's weight through its local bending.[2] After the **effect** of this primary loading has been determined, progressively turn to the other garments upon the line and consider how these supplementary **causes** will affect the preliminary sagging you just calculated for the dress. For example, the local bending required to support the blouse will **causally affect** the angle of the line as it approaches the dress from the left and **cause** you to alter the original bending under the dress accordingly. Repeat these **cause and effect** assessments over and over until you completely resolve how all of the **causal factors** operating on the dress produce a stable final configuration in which the loaded line will remain in equilibrium.

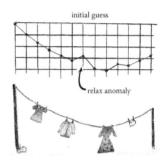

For the reasons explained in the previous chapter, a little reflection shows that the succession of "causal" stages involved in this recipe bear virtually no relationship to the temporal history through which the loaded rope will actually pass on route to the sagging curve that it eventually assumes.[3] Although I have incorporated a fair amount of "causal talk" in my account of this routine, the usage is entirely procedural in import: it merely outlines the succession of calculations a reasoner should follow in executing what Chapter 5 calls "computation by successive approximations": viz., develop a sequence of improving guesses $S_0, S_1, \ldots$ that will gradually hone in on the final clothesline curve as a "fixed point" S_∞ limit.

The old-fashioned terminology for this pattern of computation is "relaxation method" (Shaw 1953), but this label is misleading, for the component steps in its reasoning chains do not remotely reflect the friction-driven events that allow our clothesline to shed kinetic energy and permit its subsidence to a stable quiescent condition (high speed photography immediately reveals the mismatch). In fact, it is easy to see that we've not been supplied with enough data to calculate the line's

---

[2] I tacitly employ the static string equation $k\, \partial^2 y/\partial x^2 = \rho$.

[3] For some examples of misdiagnosis along these lines, see (Wilson 2006, 583–8).

true evolving history, which requires attention to both friction and the exact velocities with which the clothing items were loaded upon the cord. We merrily skirt all of these complicated contingencies by seeking its eventual static condition, which we discover by minimizing its stored energetic content through a successive approximation calculation.

The computational recipes that we follow in executing these forms of predictive reasoning are markedly different in character than we employ when we utilize a "marching method" of the sort applied to a cannonball's flight in Chapter 5.

Start with a cannonball situated at the location x_0. The governing rule $mdy/dx = g(x)$ tells us that the ball is presently angled with a slope of $g(x_0)/m$. This condition will then **cause** our ball to advance to a height close to the elevation it would reach if it retained a constant slope over the intervening displacement Δx. But as soon as we ascertain this new location, the variable gravitational factor within our modeling equa-

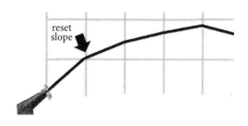

tion will **cause** the slope to alter to a somewhat different value $g(x_1)/m$. Again under the presumption that this slope will remain approximately constant over the succeeding Δx interval, compute the elevation to which this revised **causal factor** will carry our ball. And so on.

Readers may sense that "cause" is playing somewhat different roles in these two calculations, and they will be correct in thinking so. This is the theme we will now explore in the context of another model multiscalar modeling.

(iii)

As it happens, virtually all of the events we witness on an ordinary pool table can be quite effectively captured by a simple two-stage tracking technique that I shall later designate as a "hybrid modeling." It consists of two principal components: (i) an evolutionary tracking stage that plots straight line paths of the balls between their collisions and (ii) a localized supplement that covers the semi-elastic collision events themselves. The first is overly simplified because it ignores the curved trajectories that can emerge from the

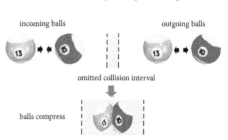

frictional interactions between the rotating balls and the nap of covering table felt. We will ignore these sophistications here.[4] The standard approach to the (ii) events was originally formulated by Wallis, Wren, and Huygens in 1668–9 and later improved upon by a supplementary "coefficient of restitution" correction that Newton added on to account for the energetic losses that inevitably arise within a two-ball encounter. This treatment is only applicable to head-on collisions, but Euler extended the basic treatment to account for ball spin and oblique encounters, adding some further "coefficients" to govern the energetic couplings between the balls. Once again, I shall ignore these supplementary refinements for simplicity's sake, as they will not affect the methodological morals I intend to draw. So I shall proceed as if Newton's simple coefficient of restitution rules are entirely adequate for this mode of modeling. Indeed, with respect to the progress of normal pool events, these rules prove extraordinarily effective, to the extent that it is hard to find pool table behaviors that lie beyond the reach of these provisos. This admirable predictability is largely due to the high standards of manufacture that govern the production of pool room equipment. If we had considered wider classes of softer or more irregular colliding bodies, lower-scale complications of the sort we will soon consider would arise with greater urgency. In his classic textbook on these subjects, Werner Goldsmith provides a nice overview of the general situation:

> The foundation for a rational description of impact phenomena was established simultaneously with the birth of the science of mechanics. The initial approach to the laws of collisions was predicated on the behavior of objects as rigid bodies, with suitable correction factors accounting for energy losses. It is interesting to note that this concept has survived essentially unchanged to the present day and represents the only exposition of impact in most texts on dynamics.... The classical theory of impact, called stereomechanics, is based primarily on the impulse-momentum law for rigid bodies and involves a minimum of mathematical difficulties in its formulation. However, it is incapable of describing the transient stresses, forces, or deformations produced, and is limited to a specification of the initial and terminal velocity states of the objects and the applied linear or angular impulse. The theory fails to account for local deformations at the contact point and further assumes that a negligible fraction of the initial kinetic energy of the system is transformed into vibrations of the colliding bodies. The last hypothesis has been found to be reasonably valid for the collision of two spheres or the impact of a sphere against a large rigid mass, but not for collisions involving a rod, beam, or thin plate.   (Goldsmith 2001, 2–4)

---

[4] There is a further reduced field of study called "mathematical billiards" which is mathematically interesting but essentially irrelevant to real-life pool table behavior (Rozikov 2019).

Although abrupt collisions of this
impulsive character were often
regarded as emblematic of how corpuscular encounters operate at
nature's most fundamental levels by
the physicists of Newton's era,[5] critics
such as Leibniz and Emilie du Châtelet
objected that this surface simplicity as
misleading, for surely the balls must

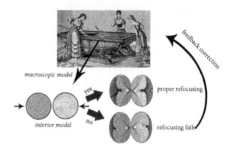

compress and re-expand over a brief temporal interval ΔT_F. During this suppressed contacting interval, stress waves must radiate from the collision point,
travel to the rear wall of each ball, reflect, and eventually refocus in a manner
that will push the balls apart with a diminished degree of vim. If the balls are
elliptically shaped like the ones that the Mikado assigns to billiard-sharps, these
rebounding waves will not focus properly, and the balls will behave very oddly.
But calculating these wave patterns requires smooth continuum mechanics in a
manner that lies far beyond the capacity of anyone unequipped with a high-
speed computer.

Although neither of them possessed any practical capacity to remedy the
situation, Leibniz's and Du Châtelet's complaints with respect to the inadequacies
of our (i) modeling seem entirely reasonable. Indeed, using a modern continuum
mechanics approximation scheme (such as a suitable finite element method), we
can attach such a direct stress wave examination as a backup submodel (ii) to our
preliminary coefficient of restitution scheme (i).

Let us pause to observe that the dissatisfactions that Leibniz and du Châtelet
express with respect to the collision-omitting tactics of (i) lie centrally in the
background of traditionalist assumptions that properly "universal principles of
causation" demand a deeper treatment of such events than (i) provides. We'll
return to this important observation later.

In even rarer circumstances, lower-scale considerations can spoil the "dominant behavior" assumptions enshrined within our (i) and (ii) submodels, requiring
a further expansion of our two-stage multiscalar scheme. For example, extremely
high transient stresses induce interfacial structural adjustments between the tiny
Bakelite crystals that compose our ball, in a manner akin to the allied processes
that can affect crystalline granite. A suitable submodel for these concerns will
utilize some generalized form of continuum mechanics that can ably handle the
so-called "configuration forces" that can reshape a material boundary (smooth
continuum mechanics cannot do this (Maugin 2011)). This is our third submodel

[5] (Wilson 2021) documents how Descartes's brand of physics (which resemble Hertz's own in many
respects) was rendered incoherent through his presumption that he should articulate its laws in two-
body collision terms.

(iii) in the chart. Finally at the truly microscopic level of the molecules that comprise the lattices within these tiny crystals, a phenomenon called "dislocation pinning" can induce behavior changes that display themselves at high-scale levels as a sudden increase in the local brittleness of the material in question. The precise details of how this transpires are complex, but suitable submodels often employ simple mass point arrangements to capture the dislocation movements (which remain invisible in schemes (i) to (iii)).

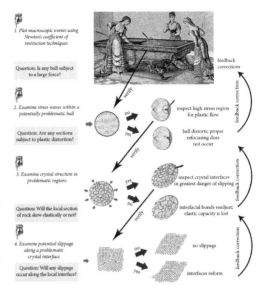

1. Plot macroscopic events using Newton's coefficient of restitution techniques

Question: Is any ball subject to a large force?

2. Examine stress waves within a potentially problematic ball

Question: Are any sections subject to plastic distortion?

3. Examine crystal structure in problematic regions

Question: Will the local section of rock skew elastically or not?

4. Examine potential slippage along a problematic crystal interface

Question: Will any slippage occur along the local interface?

feedback corrections

inspect high stress region for plastic flow

ball distorts; proper refocusing does not occur

inspect crystal interfaces in greatest danger of slipping

interfacial bonds readjust; elastic capacity is lost

no slippage

interfaces reform

Let me confess that the particular assembly of lower-scale concerns highlighted here were not selected as part of an especially cogent plan for capturing real-life billiard behavior, but because they nicely illustrate the use of the four basic classes of "classical physics" modeling policies that I placed on the four corners of our Escherian staircase, viz., (i) hybrid $\rightarrow$ (ii) smooth continuum $\rightarrow$ (iii) laminate continuum $\rightarrow$ (iv) point mass lattice. But Chapter 5 has already shown us that the "mix or match" character of this reasoning architecture will not create any difficulties as long as the component submodels confine their informational exchanges to properly crafted homogenization messages.

(iv)

We have already observed that our employments of the word "cause" in our clothesline calculations above seem rather different in character from its usage in the cannonball case. Indeed, scientists classify models like (ii) as "dynamic" in character because they track how an evolving process unfolds over time. Approach (iii), in contrast, is called "static" because it only attempts to augur how a system will resettle into an altered equilibrium condition after it has been subjected to an altered environmental condition. In the granite case, we attempt to ascertain how the rock will adjust after it has settled into a new equilibrium where it can successfully balance the increased weight of the casino. In most circumstances, we needn't worry about the transient stress waves that might ripple through the rock within the "relaxation time" interval that transpires before the rock has completed its adjustments to the added casino weight. A "static" modeling can be accordingly classified as a "revised equilibrium" modeling in which one makes the reasonable assumption that somehow or another the loaded material will find its way to a newly quiescent state that can adequately accommodate its newly acquired casino load. Much of the time these familiar "dominant behavior" assumptions prove quite trustworthy, but sometimes they are not, and a closer examination of the permanent damage that might accrue from an extreme stress wave becomes required. Potential damage of this sort is fairly common in the case of earthquakes, and a prudent engineer will need to apply some form of dynamic modeling to these transient waves to guard against dangers of this sort. Scientists normally prefer working with static models if they can, for the dangers of computational and modeling error are much higher in dynamic circumstances. Indeed, all of the four submodels used in the granite example of the previous chapter are of revised equilibrium character. We would only need to introduce a dynamic component into our scheme if we plan to erect our casino upon fault-laden soil.

But we clearly require a dynamic model at stage (ii) of our billiard scheme because the potential refocusing troubles arise within the same "relaxation time" interval Δt that our coefficient of restitution scheme collapses to 0. In contrast, submodels (iii) and (iv) remain static very much like their analogs within our granite architecture.

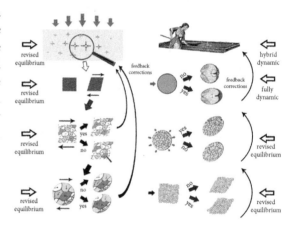

Scheme (i) can only be classified as *quasi-dynamic* due to its mixture of modeling ingredients (which is why it has been labeled as "hybrid"). Its employment within a billiard ball context has generated enormous confusion within standard philosophical discussion, for reasons we'll examine in detail later.

As it happens, applied mathematicians sharply distinguish the equation types utilized in our static and dynamic models in terms of a vital feature called their "signature" (the former employ formulas that are *elliptic* in character while the latter are *hyperbolic*). These distinctions mark significant differences in the explanatory roles that their modeling serve, and in other writings (Wilson 2017), I have attempted to explain these distinctions in laymen's terms. I will not attempt to do so here, except to remark that such considerations are extremely important with respect to a host of familiar philosophy of science dilemmas. I will merely note a few of these distinctions in behavior. (1) The hyperbolic ("dynamic") exemplars usually contain time as an independent variable whereas the elliptic ("static") models do not (time enters their dominion only as it marks the point at which a modified system has established a new equilibrium). (2) The hyperbolic collections require both initial conditions and boundary conditions before they admit of solutions as "well-posed problems," whereas the elliptic sets accept boundary conditions only. (3) This same difference explicates why we solve the two classes in markedly different manners upon a computer. In hyperbolic circum-

stances, a standard numerical method will march up a sheet of graph paper starting from the assigned initial condition in the straightforward way that we witnessed in the cannon ball case. In the left-hand illustration, I indicate how an appropri-

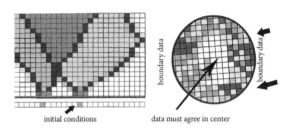

initial conditions   data must agree in center

ate "marching method" will fill out such a graph starting with a violin string plucked in two places at an initial time t_0. In a comparable static calculation (e.g., a drumhead whose bounding rim has been newly tightened), its state of internal stress needs to be calculated by working inward from the condition newly imposed along the rim. A bit of reflection indicates that finding an acceptable solution requires a large amount of provisional guesswork in the general manner of a crossword puzzle, where we are confronted with the task of verifying that our initial hunches cohere with one another as a whole. This basic need for collective coherence explains why the "elliptic problem solvers" that computer scientists employ typically incorporate a considerable degree of "successive approximation" reasoning within their routines. Physically, these repetitious recalculations are linked to the fact that for an equilibrium condition to hold within a target system, every local part of the drumhead must have managed to reach a satisfactory accord with the rest of the membrane (although as we observed with our clothesline, these two forms of "relaxation" needn't resemble each other at all).

One of my chief complaints with respect to the misleading aspects of "theory T thinking" is that its vocabularies obscure these vital distinctions by employing the standard discriminations of the applied mathematicians (such as "initial" and "boundary" condition) in extremely careless ways. *Pace* David Armstrong (Chapter 1), this policy does not provide philosophers with an "appropriate level of abstraction"; it blinds them to vital features to which they need to pay much closer attention. Modeling equations do not always operate in the same ways!

Nonetheless, I will not proceed further along these lines in this book except to remark that the intuitive differences that we detect with respect to "cause" in submodels (ii) on the one hand and (iii) and (iv) on the other directly reflect these underlying differences in equational "signature." Perhaps we have seen enough to explain my central developmental hypothesis. Beginning with a rather shapeless original usage in which distinctions of "cause" primarily mark procedural landmarks within a computational recipe, the term gradually specializes upon local features of the computational policy at hand, rather as the word "force" attaches to different physical correlates depending upon the characteristic scale length in which it is employed. In this same manner, "cause" winds up focusing upon different forms of physical consideration depending upon the dynamic or static character of the submodeling in which it appears. As this occurs, its usage sheds much of its original procedural character in favor of a more straightforward form of localized referential significance. Compared directly with one another, these discriminations will directly clash in the characteristic manner of a multiscalar architecture. Properly segregated, all of these localized uses prove descriptively useful, and none should be regarded as more fundamental than any of the others.

The "circular firing squad" character of current philosophical debate with respect to "causation" typically arises when someone attempts to favor one of these specialized sectors as clearly superior or "more fundamental" than another, a reductive proclivity that is unsuited to the underlying multiscalar character of the usage. As an illustration of the evaluative clashes that arise when these warnings are disregarded, consider the "causal" antecedents of the characteristic timbre of a plucked violin string. We might address this question in either of two familiar ways. (1) The string will quickly settle in a sustained state of characteristic vibration, determined by the final mixture of "eigenfunction" solutions to the relevant modeling equation (e.g., $c\partial^2 y/\partial x^2 = \rho$). This essentially represents a revised equilibrium form of address.[6] Such a modeling will make us inclined to view the initial pluckings as the "cause" of the resulting tonal qualities. (2) On the other hand, we may employ the same modeling equation (with its time terms retained) to plot how our initial pluckings will promulgate along the string, using an appropriate marching method for this purpose. But when we do so, we encounter a surprise, for the evolutionary modeling employed never subsides on its own into the steady vibrations we eventually expect. Instead, the little waves inserted into the string by our initial pluckings will skitter merrily back and forth along the string without any significant widening. Indeed, this is the behavior that we'll witness if we watch a high-speed capture of a suitably isolated string. We then realize that there's a lot of missing physics in our treatment and that the chief causal factor responsible for the string's subsidence into musical tonality is the product of a significant coupling between the string itself and the entire instrument's sluggish responses to the jagged wigglings of the traveling wave patterns. Moved by these somewhat surprising considerations, we may

---

[6] More exactly, it should be classified as a "steady state" problem that differs in some respects from a proper revised equilibrium modeling. However, the situations are close enough that I'll overlook these niceties here.

conclude, "The causal factor that seems most significantly responsible for the eventual timbre of the string lies within its interactions with the violin body, rather than stemming from the initial pluckings themselves." Such intervening processes are automatically overlooked when we engage in a revised equilibrium calculation because it simply presumes that a stationary final condition will somehow be eventually reached.

Under the heading of a "manipulationist approach to causation," Judea Pearl (Pearl 2009) and James Woodward (Woodward 2005) have developed a wonderfully detailed exploration of the explanatory patterns allied to our revised equilibrium considerations and their close cousins (such as steady state eventualities). Considerations of our "string and instrument body" character have led critics of this school to reject its emphases on revised equilibrium tactics as failing to fully address the underlying processes demanded within a proper account of "true causation." "Yes," it is complained, "the truncations that this school exploits are often convenient for practical purposes, but they commonly overlook essential ingredients within the full causal chain. But a proper account of 'causation' demands direct attention to the latter."

(v)

In point of fact, Mach and Poynting believe that these "fuller causal process" demands place undesirable "metaphysical" burdens upon the development of science. And they have considerable justice in doing so, for reasons to which I'll later return. Unfortunately, a coarser version of their concerns has subsequently intervened which has attracted a larger share of subsequent philosophical attention. I have in mind Bertrand Russell's well-known "On the Notion of Cause" which appears to draw directly from Mach without adequate acknowledgment (in my opinion). Here[7] Russell argues that any appeal to "causation" is otiose within modern science, because the latter generally employs differential equations that apply at single moments in time:

The law of gravitation will illustrate what occurs in any advanced science. In the motions of mutually gravitating bodies, there is nothing that can be called a cause, and nothing that can be called an effect; there is merely a formula. Certain differential equations can be found, which hold at every instant for every particle of the system, and which, given the configuration and velocities at one instant ... render the configuration at any other earlier or later instant theoretically calculable. That is to say, the configuration at any instant is a

[7] He may have reversed these opinions in (Russell 1940), for reasons that needn't detain us here. It is the quoted opinion that remains influential within contemporary circles.

function of that instant.... This statement holds throughout physics, and not only in the special case of gravitation. But there is nothing that could be properly called "cause" and nothing that could be properly called "effect" in such a system.

(Russell 1918, 194)[8]

Let us first attend to Russell's declaration that "there is nothing that can be called a cause, and nothing that can be called an effect; there is merely a formula" (he cites Newton's gravitational law as an example). However, modern mathematical practice advises that all of the descriptive ingredients employed within a well-set problem need to be considered together to properly evaluate its methodological status, not component equations considered in themselves. For example, the formula that encodes Hooke's law (stress = k x strain) does not mention passing time t, but when the formula is coupled to a suitable version of $F = ma$, we obtain Cauchy's laws for elastic response, in which the missing time "t" reappears as a second derivative and which describes wavelike causal processes of exactly the "moving forward in time" character that we discussed above.

On the other hand, when we consider a typical static modeling, the notion of "elapsed time" is directly built into the setup of the problem posed, for we want to ascertain what new equilibrium E_1 will set in after some exterior manipulation upsets an original equilibrium E_0 occurring at an earlier time. We needn't include any term for passing time within the relevant modeling equations simply because we are not interested in the transitory events required to get the system from E_0 to E_1.

But the situation is quite otherwise when we turn to dynamic models of an evolutionary character. Here Russell is quite mistaken when he declares that "there is nothing that could be properly called 'cause' and nothing that could be properly called 'effect' in such a system." In fact, there is an extensively developed discipline called "the theory of characteristics" that does exactly that, based upon the "hyperbolic signature" of the relevant

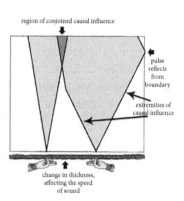

region of conjoined causal influence

pulse reflects from boundary

extremities of causal influence

change in thickness, affecting the speed of sound

---

[8] I have removed a few allusions to passages where Russell mistakenly presumes that any evolutionary model relying upon position and velocity at a single time will be fully equivalent to a similar model in which positions at two separate times are supplied. But this is simply a mistake; the latter is standardly classified as a two-point boundary problem whose existence and uniqueness characteristics differ from those of a true initial conditions problem. However, nothing material to our own concerns hinges upon this mistake, so I've simply dropped his mention of these issues.

I might mention that certain further subtleties attach to the parabolic signature equations correlated with gravitation and heat. I'm not arguing that our hyperbolic explication of "causal process" suits every pattern of successful explanation equally well, but when discrepancies arise, they merit special attention.

equations.[9] Working with these, an analyst can map out the far range of the "domains of causal influence" that emanate from disturbances at an initial time t_0 (marked in gray on the diagram), as well as the character of the effects that arise inbetween. Isn't such a narrative of evolutionary unfolding exactly what we expect from a "causal process"? But the instructions for carrying out this story are directly contained within the governing equation.

Russell also appears to endorse the thesis that causes must be distinguished from their effects by temporal criteria of some kind, lest the two notions collapse into conceptual identity.[10] Again this is simply a descriptive mistake, for causal processes driven by wavelike modelings evolve continuously from their initial conditions and don't display any evident segmentation into "cause now" / "effect later" factors. Such discriminations only appear when we convert the equation's requirements into the discrete approximation rules that we employ when we fill out a sheet of graph paper using some gappy form of marching method (i.e., calculate the elevation that a cannonball will reach if its slope $g(x_i)/m$ at x_i causes it to travel with a constant slope over the succeeding interval Δx). But this division into consecutive intervals is simply an artifact of the approximation method employed; it is not inherent in the underlying notion of a causal process itself.[11] In the *Critique of Pure Reason*, Kant criticizes Humean "before and after" thinking in exactly this vein:

> [A] room is warm while the outer air is cool. I look around for the cause and find a heated stove. Now the stove, as cause, is simultaneous with its effect, the heat of the room. Here there is no serial succession in time between cause and effect. They are simultaneous, and yet the law is valid. The great majority of efficient natural causes are simultaneous with their effects, and the sequence in time of the latter is due only to the fact that the cause cannot achieve its complete effect in one moment.   (Kant 1787, A203)

To be sure, Kant's final sentence strikes me as a lingering remnant of the tendency to confuse revised equilibrium modelings with the continuously unfolding processes we are presently considering.

---

[9] The technical distinction between these varieties of "signature" reflects the + or − manner in which component terms enter their respective differential operators. The hyperbolic opposition permits the construction of the "characteristic ones" illustrated. Any modern book on partial differential equations (Shearer and Levy 2015) explicates these matters in considerable detail.

[10] Russell's insistence that coherent causal attributions must be asymmetric in time appears to hinge upon such a Humean assumption, but this demand is completely irrelevant to the questions before us. Our wave equation is completely symmetric in its behavioral demands, but that fact doesn't prevent it from telling pulses how to march through our string in a time-forward manner.

[11] (Wilson 2017, 54–60) contains an extended critique of the manner in which these finite difference substitutions for differential equation processes have improperly encouraged logistic approaches to explanation such as Carl Hempel's familiar D-N model.

In a recent review, Hartry Field endorses Russell's central contentions as
follows. In a defense of a "physics doesn't require causation" position, Hartry
Field claims that the alleged supernumerary ingredients attaching to causal processes stem from a misleading injection of manipulationist considerations into an
otherwise cause-free scientific environment.

> I think Russell is right and [his critics are] wrong.... [N]otions like flow of
> energy-momentum and various temporal notions such as the light cone structure
> suffice for the purposes that talk of causal signals have been standardly put.
>
> (Field 2003, 435–60)

I find this comment odd because the "light cone structure" to which Field refers is
merely another label for the characteristic curves carved out by the signature of the
Maxwell vacuum equations. As such, these happen to enjoy a special status with
respect to Minkowski spacetime, but such considerations shouldn't deprive the
slower "cones of causal influence" derivable from the signatures of other hyperbolic modeling equations from enjoying the same degree of direct physical
acknowledgment.

More importantly, our current appreciation of the mathematics pertinent to
wavelike causal processes is greatly indebted to the mathematician Jacques
Hadamard (Hadamard 1903; Hadamard 1923). As it happens, various popular
"philosophical analyses of causation" bear rough affinities with the factors that
Hadamard identifies within the mathematical underpinnings of well-posed wavelike modelings. I particularly have in mind Hans Reichenbach's "mark method"
(Reichenbach 1958) and its later reworkings in Salmon (1984) and Dowe (2000). In
these accounts, Hadamard's original conception of a propagation of singularities
becomes recast as a "transfer of energy and momentum" in Field's manner,
presumably on the grounds that the latter notion seems more aptly "physicalized."
But this assimilation is simply a mistake; modeling equations of an appropriate
signature will lay down their characteristic curves and so forth whether or not
anything "material" travels along them.[12] Insofar as I'm aware, the only philosophical commentary that approaches these issues in a properly Hadamardian vein is an
excellent essay by Sheldon Smith (Smith 2000).

The long and short of these ruminations is that not all differential equations are
created alike! Russell's contentions to the contrary, a very rich collection of causal
information is directly registered within the relevant descriptive "formulas" (when
coupled to suitable side conditions), but we must ensure that we are looking at the
right kind of equations when we do so.

---

[12] (Hertz 1894, 22) offers pertinent warnings with respect to the spatial localization of potential
energy.

(vi)

But let us now consider the specific specialization of "cause" that has singlehand-edly injected the greatest degree of conceptual confusion into our understanding of the word's real-life behaviors. I have in mind a particular species of submodeling that I have largely ignored up to this point—Newton's coefficient of restitution approach. I characterized the method as *hybrid* in character because its sub-routines (and the variants that take account of the billiard table felt) employ straightforward tools to convey the balls across the table inbetween encounters. It is the treatment of collision episodes themselves that we want to investigate, for

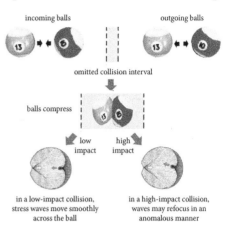

incoming balls   outgoing balls

omitted collision interval

balls compress

low   high
impact   impact

in a low-impact collision,   in a high-impact collision,
stress waves move smoothly   waves may refocus in an
across the ball   anomalous manner

Newton's tactics address these events in an abrupt manner that is often described as "impulsive." An old-fashioned terminology for a treatment of this hybrid character is "stereomechanics," but newer researches in this general vein are generally called "non-smooth mechanics" (Brogliato 2016). As we have already noted, this approach constitutes a highest-level "dominant behavior" treatment with respect to real-life billiard balls because it suppresses the rapid compressions which occasionally manifest themselves in the form of pool table anomalies. In the manner of a revised equilibrium calculation, the compressions actually arising within the balls are compressed into a "fast time" limit in a manner in which the balls are never assigned altered shapes within the (i) treatment itself, and such events are instead shuttled off in classic "lousy encyclopedia" manner to a submodeling in which suitable wave equations dynamically track the elaborate wave patterns that rapidly crisscross across the innards inwards of each ball.

As previously remarked, the degree to which everyday pool table activities can be satisfactorily handed by these simple (i) rules is truly amazing (Marlow 1995; Stronge 2004), but a large portion of this descriptive simplicity is due to diligent manufacture. The excellent degree of coherent wave refocusing exemplified within well-made balls depends significantly upon (1) their nearly perfect spherical shapes, (2) the fact that their mutual contacts initiate within a nearly point-like region, and (3) the fact that their hard Bakelite interiors can sustain significant degrees of interior stress without cracking or permanent distortion. Certainly, pool would be less fun if we pushed glass or rubber blocks about. In his classic treatise on the subject, Gaspard-Gustave Coriolis acknowledged the special empirical cutoffs that render the physics of the billiard table especially tractable to simple descriptive analysis:

> Suppose that there were a second momentum, due to elasticity, to be added to the principal momentum, due to collision during the compression of the tip. However much this second momentum might differ in direction from the main momentum, it would to that extent be impossible to play with any confidence.... [B]ut we recognize from experience that [this does not happen]. Although it is impossible for current science to establish this proposition by theory, we may nonetheless take it as sufficiently proved by experience and assume it as a basis for calculations.   (Coriolis 2005, p. 56)

But this very effectiveness has proved quite misleading, from a methodological point of view. As commonly happens, the historical progression of ideas within science and philosophy originally starts off on the wrong foot due to a misleading descriptive opportunity, for it was commonly presumed that billiard ball collisions (or something closely akin) offer our best opportunity to witness nature's processes unfolding in their most elementary form (insofar that can be possibly witnessed at a macroscopic level) (Scott 1970). This is why Wren, Huygens, Marriott, Newton, and other early modern experimenters assigned great inductive significance to events that, in reality, involve an extremely complicated array of underlying processes.[13] Indeed, such presumptions prompt Descartes into articulating his mechanical principles in impactive terms, although that packaging proves disastrous for the coherence of his views (Wilson 2022). Even to this day, philosophers frequently cite hybrid-character modelings as paragons of causal behavior—e.g., Johnny breaking a window with a rock (Paul and Hall, 2013).[14]

---

[13] Along this same front, it is amusing to note that most of the physical activities that Robert Boyle cites as admirably "explanatory" are impulsive in character:

> The next thing I shall name to recommend the Corpuscular Principle, is their great Comprehensiveness. I consider then, that the genuine and necessary effect of the sufficiently strong Motion of one part of Matter against another, is, either to drive it on in its entire bulk, or else to break or divide it into particles of determinate Motion, Figure, Size, Posture, Rest, Order, or Texture.... And as the Figure, so the Motion of one of these particles may be exceedingly diversified, not only by the determination to this or that part of the world, but by several other things, as particularly by the almost infinitely varying degrees of Celerity, by the manner of its progression with, or without, Rotation, and other modifying Circumstances.   (Boyle 1674, 141)

Any survey of the modern literature within material science quickly demonstrates how resistant to simple modeling tactics all of these behaviors have proven to be. As I've already observed, scientific investigation often begins in exactly the wrong places!

[14] The neuron diagrams favored in (Paul and Hall, 2013) are evidently of this nature, and (Salmon 1984) adopts similar assumptions (an observation for which I am indebted to Jim Woodward and Laura Ruetsche). I find Salmon's emphases especially surprising in light of the fact that his approach appears to derive (via a botched chain of transmission running through Hans Reichenbach) from Jacques Hadamard, whose important studies of causal processes were highlighted previously. But Hadamard's enlightened analyses apply to the mathematical processes that have been classified as "type (ii)" here, whereas Salmon unaccountably abandons these in favor of the considerably more problematic collection of type (i) paradigms.

Such examples encourage quite misleading conclusions with respect to the issues presently under discussion.

Insofar as I can determine, the entire subject remains adversely affected by David Hume's lingering presumption that billiard ball interactions are paradigmatic of how "causal principles" operate in their "most perfect instances":

> Here is a billiard ball lying on the table, and another ball moving towards it with rapidity. They strike; and the ball which was formerly at rest now acquires a motion. This is as perfect an instance of the relation of cause and effect as any which we know, either by sensation or reflection. Let us therefore examine it. It is evident that the two balls touched one another before the motion was communicated, and that there was no interval betwixt the shock and the motion. Contiguity in time and place is therefore a requisite circumstance to the operation of all causes. It is evident, likewise, that the motion which was the cause is prior to the motion which was the effect. Priority in time is therefore another requisite circumstance in every cause. But this is not all. Let us try any other balls of the same kind in a like situation, and we shall always find that the impulse of the one produces motion in the other. Here, therefore, is a third circumstance, viz. that of a constant conjunction betwixt the cause and effect. Every object like the cause produces always some object like the effect. Beyond these three circumstances of contiguity, priority, and constant conjunction, I can discover nothing in this cause. The first ball is in motion; touches the second; immediately the second is in motion: and when I try the experiment with the same or like balls, in the same or like circumstances, I find that upon the motion and touch of the one ball, motion always follows in the other. In whatever shape I turn this matter, and however I examine it, I can find nothing farther.   (Hume 1740, 403)

I presume that that Hume selects this particular example as a product of the popular misapprehension that Newtonian descriptive successes centrally depend upon this "billiard ball conception of nature," which is plainly incorrect (Wilson 2021). The impactive abruptness of the "constitution of restitution" treatment directly encourages the characteristic Humean theme that any attribution of causation must ultimately stem from the manner in which the film of life has been edited, which, for all we know a priori, might unfold like the disjoined montage in Buster Keaton's *Sherlock, Jr.* (Buster is a movie projection- ist who falls asleep and dreams he inhabits a world

patched together like a Hollywood film). And no inherent logical necessity attaches to how this editing is executed.

But what about our natural presumption that this attribution of abrupt response to billiard ball interaction merely arises as a convenient shortcut, and that a closer examination of detail should open out into a landscape of smoother, wavelike processes? To be sure, tracking these finer evolving details represents an exceptionally complex undertaking, impossible without modern computing equipment and highly subject to numerical error (the partial differential equations required to track the transient stresses are quite intractable and not extensively explored until much later time by Hertz (1881), inter alia). Nonetheless, these intervening events seem to render the rebounding events as *rational* in a manner that an abrupt impact approach does not. Despite Hume's claims otherwise, this request for a "rational treatment" does not seek a "necessary connection" in any modern sense of the term, but simply a larger canvas of descriptive detail.

In philosophical tradition, this explanatory preference for wavelike processes frequently takes the form of the Leibnizian adage that "nature does not make jumps," which Emilie du Châtelet succinctly characterizes as follows:

From the axiom of sufficient reason there follows yet another principle, called the *law of continuity*; it is again to M. Leibniz that we are indebted for this principle, which is one of the most fruitful in physics. It is he who teaches us that nothing happens at one jump in nature, and a being does not pass from one state to another without passing through all the different states that one can conceive of between them, ... in the same way as one does not go from one city to another without traveling along the road between the two.   (du Châtelet 1742, 133)

Du Châtelet

Leaving aside her metaphysical contentions ("the axiom of sufficient reason"), she observes that smoother behaviors are generally needed if we expect to employ the tools of the differential calculus within mathematical physics. She distinguishes the two curves illustrated by the consideration that her Fig. 3 represents a smooth curve characterized by a uniform mathematical rule whereas Fig. 4 contains an abrupt singularity lacking a tangent, in which case we lack any normal function-based rule to unite its two pieces into a rule-given unity.

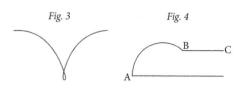

Fig. 3

Fig. 4

And these remarks embody a crucial methodological observation. Articulating a mathematical rule that can carry an evolving system through an impactive singularity generally proves quite difficult due to a lack of adequately defined bridging values. Indeed, J. C. Maxwell composed a poem on this very subject:

> Gin a body meet a body
> Flyin' through the air,
> Gin a body hit a body,
> Will it fly? and where?
> Ilka impact has its measure,
> Ne'er a ane hae I,
> Yet a' the lads they measure me,
> Or, at least, they try.   (Maxwell nd)

Although various extended forms of non-smooth mechanics can be developed, their detailed operations almost invariably piggyback upon lower-scale processes in our stress wave manner or harness associated domains of smoother behaviors to tame the singularities in the subtle manner of the "test functions" that Laurent Schwartz employs within his "theory of distributions" (Wilson 2017, 346–51). Viewing Newton's coefficient of restitution treatment of science as providing "as perfect an instance of the relation of cause and effect as any which we know" is quite off-base with respect to these reasonable methodological expectations. The successes of Newton's rules merely reflect a rather freakish form of descriptive opportunity that tells us little about the "true qualities" of causation. Insofar as I am aware, Humean sympathizers breezily ignore du Châtelet's rather natural demands upon mathematical coherence, but her observations pithily capture the dissatisfactions that critics frequently express with respect to the insufficiently elaborated details of Hume's untutored insistence upon the centrality of brute constant conjunction associations.

In passing we might observe that Hertz is fully aware of the fact that the successes of mathematical physics generally depend upon some measure of underlying smoothness, and so he excludes impactive encounters from his primal range of fundamental processes (Hertz 1894, 255).

To be sure, an overtly "metaphysical" theme is comingled within du Châtelet's remarks as well, viz., the notion that these smooth behavior requirements descend from a higher "principle of sufficient reason," a dubious claim that we shouldn't wish to endorse. This represents a palpably "metaphysical" demand upon the manner in which the physical world must behave and should be rejected as excessively strong for that reason. But her concern with avoiding unbridgeable singularities seems entirely well taken, at least at first blush.

By rearticulating du Châtelet's observations in this mathematically focused fashion, we descend from the lofty speculations of "sufficient reason" metaphysics to a smaller range of issues displaying the "even metaphysicians start small" characteristics of our opening remarks back in Chapter 1. For we are now asking a more detailed technical question, "What internal characteristics should a mathematicized description of nature display before we can regard its inferential procedures as providing a satisfactory description of the natural processes to which the usage answers?" *Prima facie*, the untamed impulsive singularities characteristic of the coefficient of restitution approach fail under this criterion.

I stress these considerations because I want to reassure my readers that I am not bluntly "anti-metaphysical" in Carnap's characteristic manner, for I regard questions of the form "how do successful patterns of mathematized description relate to the external world?" as both significant and frequently demanding of unanticipated resolutions. Many of these "small metaphysics" questions will reemerge within our survey of "scientific realism" in Chapter 8. In most cases, I will not attempt to resolve such issues definitively in this book, for I mainly wish to illustrate some of the ways in which the loose characterizations of theory T thinking and a companion emphasis upon ersatz rigor have seriously compromised our ability to diagnose and remedy the unavoidable clashes of naturally developing thinking in a satisfactory manner. In the case presently at hand, an unyielding philosophical determination to locate some "foundational" core of causal attribution generates the unhelpful spectacle of one sector of a backgrounded multiscalar architecture bickering with its neighbors for not subscribing to the same submodeling tenets. Operating from this perspective, du Châtelet's perceptive observations can be rescaled to become the accurate methodological observation that non-smooth descriptive treatments generally demand other forms of compensating submodels to fill in the gaps within their coverage. But the first step required in framing suitable observations of this "small metaphysics" character lies in detailed detective work of a Hertzian character, rather than descending from sweeping imperatives of a grander nature.[15]

Nonetheless, I am unwilling (for reason sketched in our concluding section) to acquiesce in the methodological presumption that every reasonable form of explanatory mathematical physics must someday prove reducible to a fully dynamic evolutionary tracking. At this point in time we simply do not know this to be true and must leave our understanding of what is, and what is not, feasible with respect to a mathematized approach to nature as "an altar to unknown gods" in William James's metaphor:

---

[15] In (Wilson 2017, 99–135), I argue that Leibniz's appeal to "the two kingdoms of explanation" displays a remarkably prescient recognition of the need for lower-scale corrections to upper-scale "dominant behavior" modelings.

The principle of causality, for example,—what is it but a postulate, an empty name covering simply a demand that the sequence of events shall someday manifest a deeper kind of belonging of one thing with another than the mere arbitrary juxtaposition that now phenomenally appears? It is as much an altar to an unknown god as the one that St. Paul found at Athens. All our scientific and philosophic ideals are altars to unknown gods.   (James 1884, 149)

Nonetheless, I believe that we currently know enough that we needn't concede diagnostic primacy to Humean reductionism on the grounds of the minimalist demands it places upon a worthy attribution of "causal process." It strikes me as evident that Humean advocates have pitched their tents within the least promising of our Escher staircase submodels (the hybrid corner of (i)), and du Châtelet and her allies are right to complain about the unsatisfactory qualities of the resulting treatment. This should not be regarded as stemming from any grand metaphysical pronouncement, but simply reflects a current assessment of our mathematized successes.

(vii)

But let me now add a few considerations with respect to prematurely concluding that "proper causation" irrevocably demands a supportive modeling of our class (ii), viz., as a smoothly unfolding evolutionary process. As Wilson (2017, 232–40) observes, a fair number of contemporary physicists are willing to endorse such a futuristic assertion, but I do not believe that philosophers should wholeheartedly embrace this dogmatic supposition.[16] It may well be that mathematized physics may become forced to ultimately accept a patched together façade of mathematical fragments that require resetting after a certain span of temporal or spatial separation. The notorious forms of "decoherence" seemingly demanded by quantum mechanics may support such a conclusion (Schlosshauer 2010), and the basic operations of classical thermodynamics presently rely strongly upon revised equilibrium modelings that may not prove readily eliminable as well. In this section, I will rehearse some of these latter concerns.

As it happens, Ernst Mach's strongest rationale for reducing "cause's" methodological centrality within physics as "metaphysically suspect" stems from the "energy degradations" that prove critical within equilibrium thermodynamics.[17]

---

[16] (Maudlin 2007) bases his "metaphysics" upon strong assumptions with respect to "laws" of this type, a position that I regard as prematurely presumptive.

[17] As noted earlier, (Mach 1882 and 1883) anticipate (Russell 1918) considerably, but Mach's opposition to unwanted "causal process" restrictions upon scientific development stems from a deeper range of considerations involving thermodynamics that Russell nowhere considers. Similar remarks apply to Pierre Duhem's motivations as well (Wilson 2017, 136–200).

Standard thermodynamic procedures generally utilize revised equilibrium models (like (iii) and (iv)) as their operational basis and skip over the intervening frictional events that allow the system to reach a new equilibrium. And these same gaps in mathematical coverage are naturally associated with the *energy degradation* (or "decoherence") that generally sets in when we move

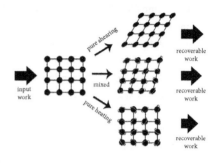

from one scale length to another within a complex material. This is because orthodox thermodynamic thinking maintains that the entropy within a material increases as coherently directed forms of inputted energy (in the form of "applied mechanical work") become progressively disorganized in a manner that prevents the original force from being regained with its full directional coherence (in the absence of outside assistance). As an illustration of this increase in energetic decoherence, subject a simple atomic lattice to an applied mechanical shear F. If all of the applied work becomes devoted to shearing the lattice in an organized pattern, our chunk of material will remain able to return this same full measure of directed force -F back to the environment (if so, the shearing is said to be *reversible*). However, this perfect form of energetic storage rarely obtains because some portion of the original F will not remain directed towards the lattice as a whole but will instead cause its component molecules to jiggle about more frenetically. As this lower-scale transfer occurs, some portion of the original F will have become degraded from a coherent directed "pressure" into a less coherent "heat."

In a complex material these degradations frequently arise within downward directed hierarchies called *energy cascades*, in which the causal connections between scale lengths ΔL divide their inputted pressures into parcels of coherent and incoherent influences. A classic illustration of a hierarchical energy cascade is displayed with the standard (if unproven) model of turbulence. If we stir water in a tub with a large paddle, a coherently oriented whirlpool initially forms on a scale length Δ_T comparable to the width of the tub. But its frictional rubbing against the surrounding water will eventually cause smaller vortices to appear at an intermediate scale Δ_I, which (like Humpty Dumpty) cannot be readily reconstituted to regain the full coherent of the parent Δ_T-scale swirling. Similarly divisive factors will

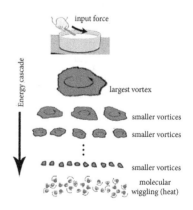

progressively degrade the rotational coherence remaining within the Δ_I vortices into yet smaller whorls.

Let me pause to again remark that none of the sample multiscale models presented above include cross-scalar thermal considerations of this type. But this merely reflects my desire for narrative simplicity; most of the structural adjustments that we witness within a real-life billiard ball or hunk of granite are highly sensitive to thermal environment. Effective forms of multiscalar modelings generally attend to these thermodynamic degradations in a significant fashion.

One of our greatest challenges in registering nature's behaviors accurately lies in capturing these decoherence processes in mathematical terms. If we attend too closely to a material's changing condition, we often fail to witness the degradation, just as we ruin the interference pattern within a two-slit experiment if we attempt to determine which slit an electron passes through. Perhaps the temporal gap within a standard revised equilibrium modeling proves a descriptive virtue, in allowing the material to shed some of its coherence unseen, rather as Clark Kent changes into Superman when he enters a phone booth.

Considered more generally, there are a variety of allied circumstances within physics where some sort of "gap" may be required to prevent a mathematized description from reaching beyond useful limits. For example, many forms of mathematized description naturally employ so-called "analytic functions" which embody an internal form of tight coherence which sometimes seems ill-suited to the physical behaviors at issue (Wilson 2006, 309–10). Sometimes we must seek some excuse for introducing a non-analytic piece into our account simply to block this analytic function zealousness from reaching too far.[18]

But do these descriptive circumstances ultimately represent an unsatisfactory state of affairs? Must all such treatments ultimately open out into fuller accounts that track in a more detailed way the manner in which the intervening "relaxation time" events unfold? I believe that a fair number of philosophers are apt to answer such questions affirmatively. Such expectations underlie the dismissive "of course" we often hear when contrary suggestions are aired. "*Of course*, we'll eventually need to inspect the suppressed degrees of internal freedom inside our billiard ball?" "*Of course*, any form of revised equilibrium treatment in thermodynamics must reduce to a fuller evolutionary tracking involving statistical mechanics?" Mach and Duhem object to these confident opinions as the products of invasive "metaphysical" prejudices that can inhibit the descriptive liberties that are desirable within science.

To be sure, both authors went further in these musings than perhaps they should, for their notorious anti-atomisms stem from the fact that they likewise

[18] (Dingle 1973, v–vi) makes the interesting suggestion that the tools commonly labeled as "asymptotic approximations" may capture a target system's behaviors more suitably than a seemingly smoother (and unrealistically "analytic") approach.

attributed these modeling preferences to our instinctive prejudices in favor of underlying "mechanical explanations":

> It is always, thus, the crude notion of substance that is slipping unnoticed into science, proving itself constantly insufficient, and ever under the necessity of being reduced to smaller and smaller world-particles.   (Mach 1882, 203)

But these unfortunate conclusions shouldn't dissuade us from recognizing the prima facie merits of not succumbing prematurely to continuous process modeling presumptions.

And certainly not on the basis of armchair "metaphysical" anticipation. A definitive resolution of these issues ultimately lies in the hands of what nature is willing to tolerate in terms of our descriptive capacities within applied mathematics, and I don't believe that it has yet supplied us with sufficient data on this score. A large number of philosophers and physicists are willing to presume that every event in the universe, large or small, must ultimately obey some fundamental set of evolutionary modeling equations, but I don't see why this is necessarily so. Successful real-life science, as we presently have it, consists in separately mathematized pieces that have cleverly been stitched together by various considerations of the sort considered in framing a successful multiscalar scheme. Perhaps all of this irregular stitching will disappear in some Utopian future, but possibly not. Insofar as I can determine, close inspection of mathematized technique reveals considerably more piecework than first meets the eye, rather than less. Even Newton himself didn't wholeheartedly trust his calculations within celestial mechanics and wondered whether angels mustn't sometimes intervene to repair some unwanted consequences within his otherwise autonomous streams of mathematized deduction.

In such circumstances, seeking a "one size fits all" analysis of "causation" strikes me as ill-suited to its probable managerial origins.[19] In view of the varying submodel tactics we have discussed, the term "cause" will naturally gravitate to

---

[19] Let me briefly mention another school of thought, with which our conclusions might be superficially confused. I have in mind the so-called "pragmatic analysis of causation" as advocated by authors such as Bas van Fraassen (van Fraassen 1980). Although pragmatic considerations of a Machian stripe certainly enter our considerations, such remarks alone scarcely qualify as a substantive account of the behaviors we have examined. To properly appreciate the conceptual tensions that attach to "cause" within scientific life, we must identify in detail the objective factors upon which the term locally focuses; the mere fact that the word displays an adaptive character is not adequately enlightening (after all, in applicational settings that involve multiple scales of length and time, most classificatory terms will display some degree of localized focusing). And it is certainly misleading to characterize these local specializations as "projections of human interests onto nature" in van Fraassen's manner. A proper evaluation operates in the opposite direction: nature "projects" its natural varieties of ΔL-scale behaviors onto the reasoning architectures we eventually learn to employ in their presence, often without realizing that our usage has been accordingly shaped by nature's "hidden hand."

rather different specimens of behavioral underpinnings. The best a philosopher might usefully offer on this score is some graduation in degrees of "causal process" involvement (similar to those offered in (Quine 1953, 139–59)). Any attempt to supply something more "metaphysically" definitive will likely trade upon the vestigial fossils of juvenile or prescientific thought, in exactly the manner of which Mach warned.

7

Dreams of a Final Theory T

> We are accustomed to call concepts metaphysical if we have forgotten
> how we reached them. One can never lose one's footing, or come into
> collision with facts, if one always keeps in view the path by which one
> has come.
>
> Ernst Mach (Mach 1909, 17)

(i)

The examples surveyed in our recent chapters suggest an attitude towards profitable usage that might be considered as a *Principle of Simultaneous Tolerance*:

> We are everywhere at liberty to match our localized selections of mathematized
> submodel to local descriptive opportunity as those opportunities become salient
> to us. All that is required of us is that we provide clear specifications on how to
> transmit corrective information from one submodel to another and to explain
> the reasons why we have adopted such a multifaceted architecture.

In doing so, I have parroted Carnap's celebrated Principle of (Absolutist)
Tolerance in its ferocious early formulation:

> *In logic, there are no morals.* Everyone is at liberty to build up his own logic, i.e.,
> his own form of language, as he wishes. All that is required of him is that, if he
> wishes to discuss it, he must state his methods clearly, and give syntactical rules
> instead of philosophical arguments.   (Carnap 1934, 52)

After his later "semantic turn," he reframed this thesis in a milder manner (Leitgeb
and Carus 2020):

> To decree dogmatic prohibitions of certain linguistic forms instead of testing
> them by their success or failure in practical use, is worse than futile; it is positively
> harmful because it may obstruct scientific progress.... Let us be cautious in
> making assertions and critical in examining them, but tolerant in permitting
> linguistic forms.   (Carnap 1947, 221)

Imitation of Rigor: An Alternative History of Analytic Philosophy. Mark Wilson, Oxford University Press. © Mark Wilson 2022.
DOI: 10.1093/oso/9780192896469.003.0007

In these codifications, Carnap displays the same concerns for permissive methodological innovation that distinguish our favored aggregation of nineteenth-century
philosopher/scientists. But he explicates these expanded liberties in a different
manner than I have suggested. Carnap's "tolerances" reflect conventionalist themes,
the presumption that different modes of syntactic usage might perform their
descriptive tasks with comparable competence, leading to a situation where any
preferential favoring of fundamental doctrine ultimately embodies a degree of
conventional choice. Many of these themes can be found in Hertz's preface, viz.,
there are a number of divergent patterns ("images") along which a self-consistent
classical physics might be developed, any of which might be freely adopted according to individual preference and convenience (Hertz is less sanguine than Carnap
with respect to the ease whereby these "observationally equivalent theories" might
be obtained). Attempting to eliminate some of these potential choices through
traditionalist philosophical strictures upon "causation" and the like represents an
ill-conceived enterprise that is likely to place unhelpful "metaphysical" restrictions
upon the radical innovations from which all of the sciences have benefited since the
beginning of the nineteenth century (Wilson 2021).

Carnap clearly presumes that Hertz is correct in this invocation of multiple
viable "images," but my own Hertz-inspired musings do not begin there. I applaud
his penetrating diagnosis of the conceptual tensions he detected within prevailing
classical mechanics practice but reject his remedy as unacceptable. Rather than
freely embracing a swatch of acceptable axiomatizations, I worry whether anything truly cogent can be obtained along these straitened corridors. So I regard
Carnap's starting point as a kind of philosophical myth to be approached with
wary suspicion. And the supplementary "convenient fiction" that these theories
can be regarded as descriptively closed unto themselves encourages the formalist
hopes that lie behind his sweeping denunciations of the metaphysical enterprise.
Carnap's axiomatic expectations rely upon an appraisal of the descriptive capacities of our mathematical tools that may well prove excessively optimistic. Our
central task in improving a worthy, yet plainly incomplete, descriptive practice
such as "classical mechanics" is to first fit our various forms of localized submodel
to the descriptive opportunities where they perform ably and then attempt to
puzzle out the strategic rationales that underlie the homogenized lines of communication we should use to interconnect these patches. These represent slowly
evolving improvements that cannot be "conveniently" transformed into completed "theories" in the mythical manner that Athena emerged fully formed
from the brow of Zeus. Like Carnap, however, I remain leery of allowing the
architectural possibilities available along these lines to be cut back through
dubious appeals to traditional metaphysical conviction. But these concerns do
not mean that smaller scale and less resolutely a priori questions do not remain
viable, even if they resist definitive resolution at this point in time. A good example
of such a concern is the issue with which the previous chapter concluded: should

we anticipate that "causal processes" of the sort exemplified within a standard "revised equilibrium" modeling will eventually become underwritten by a fuller dynamic accounting that details the transient processes leading to the new equilibrium? Or should we instead accept that abrupt jumps of a revised equilibrium character must sometimes intervene in a fashion that asks our mathematical tools to commence their descriptive tasks afresh after some episode of decoherence has transpired? This strikes me as an entirely reasonable question, to which I doubt that anyone can offer a conclusive resolution at this point in time. It also strikes me that such questions can be reasonably regarded as "metaphysical" in a "how do our words relate to the world?" manner, for they ask, "If these irreducibly singular treatments are unavoidable, what does this fact tell us about a physical world that requires these jumps?" (this is the flavor of words-to-world consideration that I characterized as "small" within our opening pages). And I fail to see such questions can be reasonably approached in any mode other than patiently attending to computational details, to our human range of reasoning capacities, and to the brute empirical successes we have reached through various experiments in strategic application.[1]

Gradualism in our descriptive prospects helps subdue the eerie sense of an external world that hides behind an impenetrable wall of theoretical posit, as we find that doctrine expressed in Hertz and Helmholtz. The adjustable architecture that interconnects the individual patches within our descriptive façades can prove just as revealing of external reality as the specific details calculated within their component submodels. We shall return to these issues under the heading of "scientific realism" in the next chapter.

The "circular firing squad" aspects of current debate with respect to "cause" seem oblivious to these considerations. If the story I have supplied with respect to the "managerial" origins of the word's developmental diversifications is proximately correct, then we should anticipate that the word's concrete employments will locally gravitate to many different forms of submodeling opportunity (I am aware of many other attractive centers of this type, but we needn't pursue these lines of complication here). Many of the major divisions in current philosophical dialectic can be roughly aligned with these basic forms of explanatory strategy. For example, the methodological emphases characteristic of the manipulationist school correspond nicely with the "revised equilibrium" tactics central within thermodynamic thinking. As we've just seen, the descriptive gaps inherent in these treatments may reflect a basic difficulty that we confront in attempting to capture the details of energetic degradation in effective mathematical terms. But self-styled

---

[1] The dedicated work of many hands will be required to resolve these issues with any assuredness, and I believe that one of the several tasks for which philosophical training renders us usefully prepared is that of preventing viable developmental pathways from being closed prematurely. All of my complaints with respect to "ersatz rigor" in this book boil down to the sense that its professionalized ideals do not encourage the veins of open-minded thinking that "being philosophical" properly requires.

"metaphysicians" frequently criticize the manipulationist approach to causation as "mere epistemology, indifferent to the notion's metaphysical grounding within the real world itself."[2] This critique clearly relies upon the raw presumption that science will eventually favor a "fundamental physics" in which all of its processes unfold in the manner of a smooth evolutionary modeling. Properly regarded, such opinions qualify as futuristic hunches with respect to our mathematical capacities that should be regarded as unresolvable at the present time. But few philosophers approach the controversies of "causation" in this undogmatic spirit, and instead presume that their disputes can be resolved through some admixture of "intuition," armchair cogitation, and a reliance upon science fiction example. But greater patience on these matters is plainly wanted, for such disagreements ultimately rest upon the question of how ably can the processes of nature be captured within uninterrupted mathematical formulas. And this last concern strikes me as one of those "small metaphysics" issues to which we presently lack a definitive answer.

Viewed through this "small metaphysics" lens, Carnap has clearly drawn his "metaphysics versus anti-metaphysics" divide in the wrong place. The proposal that analysts should "devote [them]selves entirely to the practical tasks which confront active men every day of their lives" effectively means "devote themselves to the articulation of axiomatic formalisms in the manner that Hertz hoped to find." Such an alleged expression of "tolerance" isn't truly tolerant of varied methodologies, but instead relies upon a strongly dogmatic form of unsupported metaphysical presumption (the notion that any worthy scientific proposal can be fit within the single-leveled contours of a complete and fully axiomatized coverage). As such, Carnap's allegedly "anti-metaphysical" position should be regarded as metaphysically intrusive in the same unhappy manner as our philosopher/physicists confronted with respect to the tenets demanded by the philosophical schools of their time.

Oddly enough, present-day divisions between self-styled "metaphysicians" and their "anti-metaphysical" opponents remain roughly where Carnap left matters (Chalmers, Manley, and Wasserman 2009). I hope that our emphasis upon the "small metaphysics" underpinnings of these divisions indicates that plenty of daylight lies between these two extremes.

(ii)

A second unfortunate emphasis that we owe to the "logical empiricists" of Carnap's era lies in their propensity to what might be called the "logicization of investigative terminology" (these represent the diagnostic predilections that I usually call "theory T thinking"). As I earlier acknowledged, the formal

---

[2] Woodward (2021) offers an excellent survey of popular "strategies of dismissal" of this general character. But one can't carve out a coherent discipline on the grounds that one happens to have a pre-existent name for it (such as "astrology").

codification of first-order logic in the twentieth century has undeniably produced a greatly improved understanding of the notion of "deductive consequence" and its salience with respect to the relationships that a theorem should rigorously bear to a generating set of initial axioms. But as frequently occurs with the discovery of any new tool, enthusiasts immediately apply the technique to problems to which it is not particularly well suited. In this vein, authors like Carnap enthusiastically proclaimed that the nineteenth-century struggles over methodology could be swiftly and cleanly dispatched with the assistance of this new logicized formalism. But for this program to prove viable, real-life scientific theories must submit to comprehensive axiomatization, that can regiment their special vocabularies by some form of "implicit definability." This was the sweeping prospect that entranced Carnap, who proudly proclaimed in 1931 that the advances of "applied logic" provide a universal recipe for ridding science of unproductive metaphysical intrusions of which predecessors like Mach had rightfully complained:

> The development of modern logic has made it possible to give a new and sharper answer to the question of the validity and justification of metaphysics. The researches of applied logic or the theory of knowledge, which aim at clarifying the cognitive content of scientific statements and thereby the meanings of the terms that occur in the statements, by means of logical analysis, lead to a positive and to a negative result. The positive result is worked out in the domain of empirical science; the various concepts of the various branches of science are clarified; their formal-logical and epistemological connections are made explicit. In the domain of metaphysics, including all philosophy of value and normative theory, logical analysis yields the negative result that the alleged statements in this domain are entirely meaningless.... [I]t is only now [that] the development of logic during recent decades provides us with a sufficiently sharp tool that the decisive step can be taken.   (Carnap 1932, 60–1)

But this merely represents a fond hope, not a truly feasible program for helpful conceptual analysis, as we have learned from Hertz's failure to resolve his original conceptual confusions through axiomatic means. But the prospect of a tidy dismissal of all philosophical squabbles appealed so strongly to Carnap and his friends that the policy of *pretending* that "scientific theories" spring fully formed axiomatically from the brows of their creators became widely accepted as a convenient "epistemological idealization" able to "exhibit the heart of many philosophical problems... disentangled from confusing empirical detail" (to quote David Armstrong from Chapter 1 once again). In acquiescing in these myths of immaculate conception, gullible philosophers have detached themselves from the symptomatic irregularities that suggest that an effective descriptive scheme may not encode its physical data in any simple variety of "Fido"/Fido pattern, leading to the "small metaphysics" concerns with respect to applied mathematics' descriptive reach that we have just outlined.

The standards of "ersatz rigor" of which this book complains reflect this excessive reliance upon "epistemologically idealized theories," which inaccurately presume that the troublesome anomalies of real-life descriptive practice will someday become vanquished with a swift sweep of the axiomatic pen. In doing so, excessive attention to the ersatz rigor of surface-logicized order obscures the need to scrutinize the crooked details of real-life application with a critical eye. The latter typically demands a "rigorous examination" of clues in the manner of a scrupulous detective, but not in a fashion where we already know the plot ahead of time.[3]

Carnap's asseverations otherwise, standard logical discriminations do not offer "sharp diagnostic tools" with respect to a judicious appreciation of methodological variety. Consider this characteristic appeal to an alleged distinction between "laws" and "initial conditions" (from Ernst Nagel):

> The first point to note in this example is that the premises contain a statement that is universal in form and that asserts an invariable connection of certain properties.... If now we generalize from this example, it appears that at least one of the premises in a deductive explanation of a singular explicandum must be a universal law, and one moreover that is not a sleeping partner but plays a role in the derivation of the explicandum.... But in addition to a universal law, the above premises also contain a number of singular or instantial statements, which assert that certain events have occurred at indicated times and places or that given objects have definite properties. Such singular statements will be referred to as ... "initial conditions." ... The indispensability of initial conditions for the deductive explanation of individual occurrences is obvious as a point in formal logic.... Accordingly, a deductive scientific explanation, whose explicandum is the occurrence of some event or the possession of some property by a given object, must satisfy two logical conditions. The premises must contain at least one universal law, whose inclusion in the premises is essential for the deduction of the explicandum. And the premises must also contain a suitable number of initial conditions.   (Nagel 1961, 30–2)

Terminologies like "initial condition" and (to a lesser extent, "law")[4] are extracted from applied mathematics and whimsically adopted here, with a considerable loss in clarity (I find it astonishing that many knowledgably parties such as Nagel have

[3] To be sure, Wilson (2006) discusses the typical "seasonalities" that attach to the decoding of a practical usage that is initially ill-understood; we should distinguish the preliminary diagnostic work emphasized here from the subsequent reformations of inferential procedure we may reach based upon a revised "semantic picture." The latter task generally benefits from the close attention to data transmission characteristic of a so-called "correctness" or "soundness proof." In this second enterprise, formal rigor of a logistical character usually proves quite important. My chief claim in this book is that these eventual recastings should be evaluated in Oliver Heaviside's manner: as the tails that eventually complete an improved dog, not as the core canine itself. Our multiscalar architectures indicate such studies of "data transmission" can be aptly applied to topics that are not axiomatically arranged.

[4] In particular, passages like Nagel's muddy the vital distinction between what Woodward and Wilson (2019) call "system laws" and more reasonable candidates for the traditional label "law of nature."

been willing to mistreat good words so). Acting as if the distinction merely reflects a "point in formal logic" immediately erects severe barriers to clear thinking. At the cost of a mild digression, let me quickly cite an example from some of my current research work. Long ago Lord Rayleigh complained of what might look like an inconvenient technicality:

> [I]t may be well to emphasize that a simple vibration implies infinite continuance and does not admit of variations of phase or amplitude. To suppose, as is sometimes done in optical speculations, that a train of simple waves may begin at a given epoch, continue for a certain time involving it may be a large number of periods, and ultimately cease, is a contradiction in terms.   (Rayleigh 1900, 486)

As often happens, this annoying "technicality" has proved to be the specific thread that has subsequently unwoven the entire sweater of what we now call "classical optics," leading to the surprising "I've tried not to lie to you too much up to now" passages that one finds halfway through an advanced textbook:

> Any light that isn't perfectly monochromatic (and that's every kind of light) necessarily has some degree of randomness. In Chapter 9 we described this randomness... [but] there were obvious inadequacies; having no quantity to represent a "degree of coherence"; we could not give a precise condition for coherence to disappear or for fringes to be so indistinct that they should be considered absent. Moreover, we found ourselves using awkward wordings: "obtaining knowledge"; "permits us to predict"; "so clumsy that interference fails to manifest itself"; "a wavetrain is the whole wave defined for all t."   (Brooker 2003, 219)

At the core of these complexities is the fact that articulating a notion of "initial condition" suitable to light is quite subtle, because a so-called "microlocal" articulation is required that can allow a concentrated pulse to radiate in different directions when it enters a refractive medium (these circumstances demand what Berkeley might have called "a ghost of a departed frequency"). The upshot reminds me of the old English broadsheet, "The World Turned Upside Down" because resulting reanalysis leaves most of the familiar highlights of optical research firmly in place, yet with dramatical adjustments in their semantic support. We will consider allied reappraisals of "support" in the next chapter.[5]

THE
World turn'd upſide down:
OR,
A briefe deſcription of the ridiculous Faſhions
of theſe diſtracted Times.
By T.J. a well-willer to King, Parliament and Kingdom.

London : Printed for John Smith. 1647.

---

[5] I am presently working on a report provisionally entitled "How 'Wavelength' Found its Truth-Values."

This same "logification" of applied mathematical classification cuts up a smooth trajectory into the stepwise "A obtains at time t" / "B obtains at time t + Δt" transitions beloved of Humeans. I've already remarked on the unhelpful aspects of such an approach.[6]

(iii)

So much for the typical fashion in which excessively militant "anti-metaphysicians" of a Carnapian stripe covertly rely upon fictitious forms of theory T structure. But what about the other side of the current coin, the cohort of analytic philosophers who have recently embraced the "metaphysics" label in a more robust manner. Oddly enough, their efforts also remain entrapped within the clutches of theory T thinking without their fully realizing that fact.

Earlier in our discussion we noted that Hertz and Mach strongly dissent from a traditional approach to conceptual clarification in which the analyst opens up our suppressed "receptacles" of word meaning to locate the hidden ingredients responsible for the surface difficulties we confront:

> When we examine what actually goes on in our minds when we are doing intellectual work, we find that it is by no means always the case that a thought is present to our consciousness which is clear in all its parts. For example, when we use the word "integral", are we always conscious of everything appertaining to its sense? I believe that this is only seldom the case. . . . Our minds are simply not comprehensive enough. We often need to use a sign with which we associate a very complex sense. Such a sign seems, so to speak, a receptacle for the sense, so that we can carry it with us, while being aware that we can open this receptacle should we have a need of what it contains.   (Frege 1914, 209)

But Hertz strongly insists if you search for conceptual disharmonies in this introspective fashion, you'll usually come up empty, for the tensions lie in the broader fabric of our applicational policies, not localized within individual words. Carnap, of course, subscribes to the same moral, but one would prima facie anticipate that our new metaphysicians would claim to find the underpinnings of their metaphysical strictures upon "causation" buried within the word in a traditionalist manner. Oddly enough, in my experience they become quite offended if one attempts to saddle them with such manifestly old-fashioned "analytic philosophy" presumptions.

---

[6] Wilson (2017, 54–63) surveys the hidden role that finite difference approximations play in the appeal of Hempel's (1970) popular "D-N model of explanation."

Instead, they commonly contend that they seek "theories" of metaphysical tenet that capture the intuitive demands that scientists follow in their later excursions into empirical science.[7] In doing so, these philosophers believe that they follow the same methodological imperatives of their scientific colleagues except that philosophers attempt to organize our prescientific conceptual demands. In this vein, L. A. Paul claims that, despite the fact that "many concepts of metaphysics are conceptually prior to the concepts of science," their anticipated treatments closely resemble one another methodologically:

> We can theorize about the world using models, that is, by constructing representations of the world, and metaphysical theorizing is no exception. Scientific theorizing is often understood in terms of the construction of models of the world, and scientific theories about the nature of features of the world may be understood as models of features of the world. Metaphysical theories about the nature of the features of the world may also be understood as models of features of the world. Both fields can be understood as relying on modeling to develop and defend theories, and both use a priori reasoning to infer to the best explanation and to choose between empirical equivalents. On this view, the most important differences between the scientific method and the metaphysical method derive merely from difference in subject matter and the resultant difference in the role they give ordinary experience.   (Paul 2012, 6 and 90)

In such a pursuit, "philosophical models" can be proposed that are "idealized" in the manner of the "frictionless planes" of the physicist, without serious concern for the occasional "intuitive" counterexample. And the stringent demands of "conceptual analysis" in Frege's "open the receptacles of thought" vein can likewise be avoided, without needing to closely examine concrete scientific practice in the nitty gritty manner I have recommended. Such methodological opinions are significantly influenced by (Lewis 1983).

In this same vein, Theodore Sider presumes that these pre-scientific arrangements supply the conceptual slots that concrete science later fills with suitable "natural properties":

> A realistic picture of science leaves room for a metaphysics tempered by humility. Just like scientists, metaphysicians begin with observations, albeit quite mundane ones: there are objects, these objects have properties, they last over time, and so on. And just like scientists, metaphysicians go on to construct general theories based upon those observations, even though the observations do not logically settle which theory is correct. In doing so, metaphysicians use standards for

---

[7] Such "we propose theories just like everyone else" defenses of the philosophical enterprise are extremely popular nowadays (Daly 2010, 155–84).

choosing theories that are like the standards used by scientists (simplicity, comprehensiveness, elegance, and so on).   (Sider 2007, 6)

From these points of view, a metaphysical theory of causation belongs to what might be called "the prescience of science": the basic conceptual repertory we must allegedly master before we become ready to carry out the parochial objectives that define genuine science thereafter. In the fuller account of these "mundane observations" that is provided within Sider (2014), he extends his pre-scientific categories to embrace most of the discriminations advanced within the standard taxonomies of theory T thinking (e.g., "projectible predicate," "law of nature," "initial and boundary conditions," "cause," and so on). The anointed task of a Sider-style metaphysician becomes one of determining how these critical categories relate to one another—"Can 'law' be reduced to Humean regularity?," "Can 'cause' be reduced to 'law'?" Or *vice versa*? Because this self-styled "prescience" only deals with the organizational categories that every permissible concrete "science" must exemplify, Sider presumes that these studies can be largely conducted on an armchair basis, without needing to delve into the messy particularities of "force" and "constraint" in the manner we have considered.

(iv)

Such opinions fail to address the questions that Mach piquantly captures in his writings on mechanics, "Where do our intuitions on such matters come from and how firmly should they constrain the future improvements of science?" Mach warmly acknowledges the guiding virtues of intuitive hunches but warns against "creating a new mysticism out of the instinctive":

It is perfectly certain, that the union of the strongest instinct with the greatest power of abstract formulation alone constitutes the great natural inquirer. This by no means compels us, however, to create a new mysticism out of the instinctive in science and to regard this factor as infallible. That it is not infallible, we very easily discover. Even instinctive knowledge of so great logical force as the principle of symmetry employed by Archimedes, may lead us astray. The instinctive is just as fallible as the distinctly conscious. Its only value is in provinces with which we are very familiar.... Still, great as the importance of instinctive knowledge may be for discovery, we must

Mach

not, from our point of view, rest content with the recognition of its authority. We must inquire, on the contrary: Under what conditions could the instinctive knowledge in question have originated? We then ordinarily find that the very principle to establish for which we had recourse to instinctive knowledge, constitutes in its turn the fundamental condition of the origin of that know-ledge. And this is quite obvious and natural. Our instinctive knowledge leads us to the principle which explains that knowledge itself, and which is in its turn also corroborated by the existence of that knowledge, which is a separate fact by itself. (Mach 1883, 34–5)[8]

By the latter, Mach intends the descriptive opportunities that have rewarded specific strategies of linguistic exploitation (revised equilibrium modelings, for example) and asks us to beware of presuming that similar patterns of thought are required within every branch of descriptive endeavor. But the latter represent the dangers that converting useful propensities into metaphysical verities pose. If we bear in mind the evolutionary manner in which classificatory tactics locally condense from a more amorphous magma, we will remain less inclined to confuse the intuitive with the obligatory. As Mach remarks in the epigraph to this chapter:

We are accustomed to call concepts "metaphysical" if we have forgotten how we reached them. One can never lose one's footing, or come into collision with facts, if one always keeps in view the path by which one has come.

The ersatz rigor notion that a suitable "theory" of intuitions should prove intern-ally coherent in a quasi-axiomatic manner plays a role in this lack of reflection with respect to the origins of our armchair predilections. Our experience with other classificatory words suggests that original inclinations will eventually become diversified in a multiscalar manner.

With respect to this last point, we observed that the scientific investigation of "idealized models" is usually accompanied by the expectation that their dominant behavior predictions will be someday tempered by a surrounding string of sub-model corrections. They should not be regarded as entirely acceptable construc-tions unto themselves.

---

[8] See also Helmholtz:

[T]his is however in disagreement with the older concept of intuition, which only acknowledges something to be given through intuition if its representation enters consciousness at once with the sense impression, and without deliberation and effort....I believe the resolution of the concept of intuition into the elementary processes of thought as the most essential advance in the recent period. (Helmholtz 1878, 130 and 143)

I have subtitled this volume as an "alternative history of analytic philosophy," because it attempts to sketch a trajectory that our subject might have followed had it attended more warmly to the diagnostic observations of our nineteenth century philosopher/scientist forebears with respect to the gradual stabilization of an originally amorphous descriptive practice. We have instead become distracted by the false allure of self-enclosed axiomatic formalisms and have neglected to attend to our grubby detective duties.[9] The resulting disdain for the unexpected turns of unfolding linguistic process as "merely epistemological" and irrelevant to the "deeper metaphysics" of the external world strikes me as pound-foolish in the extreme. The significant confusions with respect to "force" and "cause" that this book has documented take their origins within the convoluted pathways of natural linguistic development and cannot be readily severed from them. S. T. Coleridge observes:

> For language is the armory of the human mind; and at once contains the trophies of its past, and the weapons of its future conquests.   (Coleridge 1817, II, 29)

---

[9] It should be remarked that none of our favorite philosopher/physicists were willing to extend the same open-minded interpretative policies ("scientific realist" in the sense of Chapter 8) to mental classifications such as "I am experiencing a red sense datum." Instead, they clung to the doctrine of immediate acquaintance that is commonly dubbed as "the incorrigibility of the mental." Indeed, Hertz's contention that "we form for ourselves images or symbols of external objects" wouldn't make coherent sense otherwise (Wilson 2006, 390–401).

8

Linguistic Scaffolding and Scientific Realism

A deformable body, as distinguished from a rigid body, is one which is susceptible of a change of form either as a whole or in any of its parts. Strictly speaking, all bodies in Nature are deformable, for there are no motions which are not accompanied by greater or smaller changes of form in the bodies participating in them. But in many cases—for example, in that of the pendulum, the lever, the top—it is sufficient, for a first approximation to reality, to assume the bodies in question as rigid....Nature does not allow herself to be exhaustively expressed in human thought.

Max Planck (Planck 1932, 1–2)

(i)

At a certain point in his discussion, Heinrich Hertz explicitly disassociates his own philosophical opinions from the more militant forms of anti-metaphysical opposition that he saw in fellow philosopher/physicists such as Ernst Mach:

The usual answer, which physics nowadays keeps ready for such attacks, is that these considerations are based upon metaphysical assumptions; that physics has renounced these, and no longer recognizes it as its duty to meet the demands of metaphysics....If we had to decide upon such a matter, we should not think it unfair to place ourselves rather on the side of the attack than of the defense. A doubt which makes an impression on our mind cannot be removed by calling it metaphysical; every thoughtful mind as such has needs which scientific men are accustomed to denote as metaphysical. Moreover, in the case in question, as indeed in all others, it is possible to show what are the sound and just sources of our needs.   (Hertz 1894, 23)

Hertz does not amplify his concerns fully, but he appears to have in mind some vital "small metaphysics" considerations that have become almost completely obliterated within the cruder, post-Carnapian conception of how an opposition

Imitation of Rigor: An Alternative History of Analytic Philosophy. Mark Wilson, Oxford University Press. © Mark Wilson 2022.
DOI: 10.1093/oso/9780192896469.003.0008

to "metaphysics" should be framed.[1] My central purpose in this penultimate chapter is to elevate these suppressed "small" concerns to philosophical attention once again. Perhaps surprisingly, Hertz begins his discussion by declaring that the first part of *Principles* should be regarded as synthetic a priori in a Kantian vein throughout, with no direct reliance upon empirical behavior at all:

> The subject-matter of [our] first book is completely independent of experience. All the assertions made are a priori judgments in Kant's sense. They are based upon the laws of the internal intuition of, and upon the logical forms followed by, the person who makes the assertions; with his external experience they have no other connection than these intuitions and forms may have.   (Hertz 1894, 45)

These remarks are notoriously cryptic (Lützen 2005, 121–3), but Hertz clearly hopes to assemble a specific realm of firm objective data to which physics should be held responsible. In doing so, we should follow the reasoning procedures which strike us as proving the best guarantors of deductive soundness, which Hertz identifies with guiding principles of traditional Euclidean geometry (in company with Kant). But this merely represents a jumping off point, for Hertz also embraces a two-tiered methodological approach in which enterprising physicists remain free to follow their creative hunches into significantly distinct "extension realms," like the projective realm that had been devised by projective geometers earlier in the century (in which Euclid became supplemented with large quantities of totally unintuitive "extension points") (Wilson 2020). Hertz himself engages in a large amount of rather outrageous supplementation of this general character, as outlined in the appendix to Chapter 3.[2] But it is only within this enhanced realm that Hertz finds the platform upon which his hedged form of scientific realism ultimately rests, for "the totality of things visible and tangible do not form a

---

[1] The specific "metaphysical" objection he raises applies to employing Hamilton's principle as a "second image" foundational doctrine on the grounds that it "attributes intentions to inanimate nature" (Hertz 1894, 23). As indicated previously, he expands this criticism by pointing out that such tactics pick out the wrong behaviors in the case of non-holonomic conditions (choosing paths of optimal control rather than the proper frictionless response—see Bloch (2003)). Hertz's discussion here is carefully and admirably observed, and I have been surprised to find later commentators loosely presuming that Hertz regards his three "images" as descriptively equivalent. But it is also common for contemporary philosophers to loosely conclude that various physical doctrines are "fully equivalent" to one another on the grounds that they coincide within a purist point mass setting of the "Newtonian" sort discussed in Chapter 3.

[2] He appears to assume that we should follow Kant's Euclidean geometrical percepts in constructing the base-level properties of the interlocked mechanisms we find in nature, viz., the ways in which their parts connect with one another and the freedoms of movement they display as a connected ensemble. These geometrical constructions then characterize the "mobility space" facts that comprise the prime data to which Hertz's system of mechanics is ultimately responsible. The reliability of these constructions is fully guaranteed by the a priori principles we employ in ascertaining these basic mobilities. However, to frame a universe governed by simple and comprehensive laws. we must enrich these base-level descriptions into becoming the extended universes described in the appendix to Chapter 3 (i.e., as the manifolds $\mathcal{M}$ that cover the mobility spaces S).

universe conformable to law" (Hertz 1894, 25–6). Indeed, the empirical "law" that ultimately governs dynamic change within Hertz's hypothetically enlarged realm appeals to movements within a high dimensional manifold bearing a metric of a decidedly non-Euclidean character. Hertz clearly believes that a properly conceived "metaphysics" should not prohibit the ingenious scientist from proposing radical reappraisals of this type.[3]

This is where Hertz's appeal to a parallelism between "images in thought" and "consequences in nature" enters the scene; it provides the bridge whereby he reaches out to the external world from the framework assembled in the *Principles*. He acquired this minimalist recipe for learning about "things in themselves" from his teacher Helmholtz, who wrote in a celebrated passage:

> Natural science...seeks to separate off that which is definition, symbolism, representational form or hypothesis, in order to have left over unalloyed what belongs to the world of actuality whose laws it seeks.... The relation between the two of them is restricted to the fact that like objects exerting an influence under like circumstances evoke like signs, and that therefore unlike signs always correspond to unlike influences. To popular opinion, which accepts in good faith that the images which our senses give us of things are wholly true, this residue of similarity acknowledged by us may seem very trivial. In fact, it is not trivial. For with it one can still achieve something of the greatest importance, namely forming an image of lawfulness in the processes of the actual world. Every law of nature asserts that upon preconditions alike in a certain respect, there always follow consequences that are alike in a certain other respect. Since like things are indicated in our world of sensation by like signs, an equally regular sequence will also correspond in the domain of our sensations to the sequence of like effects by law of nature upon like causes.... Thus although our sensations, as regards their quality, are only signs whose particular character depends wholly upon our own makeup, they are still not to be dismissed as a mere semblance, but they are precisely signs of something, be it something existing or happening, and—what is most important—they can form for us an image of the law of this thing which is happening.   (Helmholtz 1878, 118)

As we noted in our Chapter 5 criticisms of Hertz's "images in thought/ consequences in nature" passage, these simple parallelisms are rarely valid, and the facts of objective nature must be determined from our pragmatically rewarded procedures in other ways, in a manner to which I will later return.

---

[3] Scientific realists are generally inclined to reject the "world of familiar acquaintance" themes of Stebbing (1937) (see also Wilson (2017, 40–6)). It was common in Hertz's era to presume that free-spirited human creativity represents a mental capacity superior to the allegedly "constant conjunction" character of animal thinking (Wilson 2020, 66–7). Whether Hertz subscribes to such themes or not, I cannot certify.

We should observe that such doctrines rely heavily upon the traditional presumption that we are intimately and fully acquainted with the evanescent qualities of our mental experiences (Hertz's "images"), but the internal feature of the objective properties of the external world remain permanently beyond our intellectual grasp. Within this wintery assessment, all we can ever learn about the external world is supplied within the behavioral husks that we can glean from the Harpo-imitates-Groucho parallelisms upon which science depends. This dichotomy between our allegedly robust "grasp" on the "mental" and our feeble acquaintance with the "physical" remains with us to this day, although the better angels of twentieth-century philosophy vigorously attempted to reject these unfortunate motifs of "private language" thinking. I will merely comment that I believe that these critics are right.

(ii)

However we correct these misguided doctrines with respect to conceptual "grasp," a fundamental quandary persists. To frame a cogent portrait of external circumstance, we must stoutly rely upon the principles of mathematical reasoning that we regard as inherently the most sound and reliable. But these same reasoning tools may incorporate interior ingredients that we will not wish to foist upon the external world. Hertz's appeals to the securities of Euclidean construction apparently trace to the fact that these tactics undergird our firmest advances into the inferential unknown. But they frequently do so in a manner that may strike us as rather mysterious from a psychological point of view. Consider the Pythagorean theorem. Staring blankly at a right triangle, we may idly frame the hunch that the squares on its sides add up in a Pythagorean manner, but this is not an ironclad proof. But imbed this figure within an appropriate nest of supplementary lines, and our triangle becomes entrapped within a mathematical box that leaves 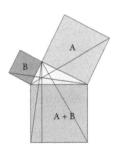 the figure no option except to prove Pythagorean. Clearly deep mathematics gains many of its ampliative powers through more radical forms of supplementary embedding. Within an unanticipated richer setting, internal relationships that had heretofore escaped our notice may become vibrantly salient (Kant characterizes this surprising source of supplementary certainty as the "synthetic a priori"). Unlike Mach, Hertz does not want to adopt raw sense data as the base domain to which science is ultimately responsible, but to instead utilize the geometricized constructions that Kant tells us how to construct via Euclidean principles (it is here that we obtain the machine-like "mobilities" discussed in Chapter 4).

But this is where a new dilemma intrudes. Given Hertz's strong reliance upon intermediate ampliative reasoning of this Euclidean character, how can we coherently understand his contention that, for all that, the external world itself might be credited with a geometry of a markedly different character (or, possibly, no certifiable geometry at all)? A few years after Hertz's *Principles*, the physicist J. H. Poynting evocatively captures these methodological concerns as follows:

> While the building of nature is growing spontaneously from within, the model of it we seek to construct in our descriptive science, can only be constructed by means of scaffolding from without, a scaffolding of hypotheses. While in the real building all is continuous, in our model there are detached parts, which must be connected with the rest by temporary ladders and passages, or which must be supported till we can see how to fill in the understructure. To give the hypotheses equal validity with the facts is to confuse the temporary scaffolding with the building itself.
>
> (Poynting 1899, 385)

Poynting

Some of the "temporary scaffoldings" that Poynting appears to have in mind are the roundabout but seemingly necessary "foundational" appeals to point masses or rigid bodies to which Victorian physicists first appealed in their attempts to reach an exterior universe comprised of fields and the other forms of continuous media capable of supporting wavelike processes. In the appendix to Chapter 3, I outline the manner in which these nineteenth century appeals to "essential idealization" were later addressed with the direct mathematics of limits and integral equations (these replacement pathways were suggested by Hilbert in his 6th problem on mechanics). Despite these improvements, the residual worries of a Poynting remain: in framing our portraits of the external world in mathematical terms, at which point do we become free to throw away the unwanted features that may accompany these constructions? Joseph Boussinesq writes, in advancing a standard Kantian theme:

> The human mind while observing natural phenomenon recognizes in them, besides many clear elements which it does not get to unravel, one clear element, which by virtue of its precision is liable to be the object of truly scientific knowledge. That element is the geometric one, pertaining to the localization of objects in space which permits one to represent them, to draw them, or to construct them in a more or less ideal manner.   (Boussinesq 1880, 88)

But if we don't regard these human "precisions" as fully representative of external reality, how should we proceed? This is exactly the dilemma that Hertz confronts in his two-tiered approach to satisfactory modeling (viz., he claims that all of the conclusions reached in the first half of *Principles* are thoroughly Euclidean in their mathematical reliability, yet in his "empirical" follow up he abandons that preliminary scaffolding in favor of higher-dimensional manifolds possessing a non-Euclidean metric). But it is only at this second stage of construction that Hertz finds the altered stream of mathematical "images" that he hopes can successfully parallel the objective structures present in external reality. But if we must first climb up a Euclidian ladder to reach a decidedly non-Euclidean perch, upon what basis are we allowed to throw our initial scaffolding away in reaching a truer estimation of the strange new world that lies beyond?

> Scientific accuracy requires of us that we should in no wise confuse the simple and homely figure, as it is presented to us by nature, with the gay garment which we use to clothe it. Of our own free will, we can make no change whatever in the form of the one, but the cut and color of the other we can choose as we please.
>
> (Hertz 1893, 28)

At this point a number of "small metaphysics" questions pertaining to the reliability of our mathematical capacities make their entrance. What degree of similitude can we expect will hold between the streams of calculation that we are able to certify as mathematically rigorous and the behavioral expectations that we hope to attribute on this basis to the world beyond?

Mach's brutal response to Poynting's worries with respect to excessive scaffolding maintains that we needn't worry about any hypothesized external world at all, beyond a pure phenomenalism of a neutral monist complexion. Like most of us, Hertz would prefer avoiding these anti-realist extremes without embracing any of the otherwise intrusive metaphysical demands that Mach rightly rejects. Hertz accordingly decides that some residual dose of smaller-bore "metaphysics" is needed to address Poynting's concerns with respect to "scaffolding." And my own assessment of our intellectual circumstances remains roughly the same.

What is the striking fact about Kantian thought that appears to have particularly impressed Hertz? Most likely it resides in the ampliative certitude that we mysteriously gain from following the strict constructive pathways that Euclid outlines. Even if we eventually decide to extend our constructions beyond this point in Hertz's "second part" manner, we recognize that their descriptive merits derive entirely from the ways in which they simplify the Euclidean basis upon which they are built. But similar crablike advancements in inferential trustworthiness affect every portion of effective applied mathematics. Euler mistrusted Daniel Bernoulli's hunch that arbitrary positions of a violin string could be spectrally decomposed into sine wave factors, but this factorization capacity was mathematically ratified later by Fourier, Dirichlet, and Sturm and Liouville (thereby supplying nineteenth-century physics with the vital "poem" (Thomson and Tait 1912, 470)) that underwrote many of the subsequent advancements achieved in the over half century that follows. But the security of Sturm and Liouville's advancements is ratified only by embedding the target problems within a set theoretical netting of a decidedly modernist cast (Wilson 2017, 399–402).

The stepwise character of intellectual advance stems directly from our inherently limited reasoning skills that we must somehow transcend through clever stratagems and wiles. We merely need to remember, as Mach constantly stresses, the fact that evolutionary process has fashioned us with computational capacities better suited to the homey tasks of securing food and shelter than unraveling the exterior mysteries of light and matter. We collectively face a methodological dilemma similar to the one that Descartes confronts in his *Sixth Meditation* (Descartes 1641): how can we ascertain the true features of the exterior world, given that we must largely depend upon sensory signals that have been overtly fashioned to keep our bodies safe and healthy rather than becoming objectively knowledgeable? Descartes's striking answer appeals to reverse-engineering the reasons why we possess the sensory systems we do, by considering how limited creatures such as ourselves might plausibly deal with the complicated environments in which we find ourselves. From this perspective we find that the sensory signals we intuitively regard as "simple" actually represent highly processed responses[4] to external promptings that allow human beings to address their needs in swift and generally effective ways. The better we can appreciate the reasons why our sensory skills are constructed in this biased-in-favor-of-human-capacity manner, the more clearly we will recognize what the objective world is like in its own right. These Cartesian doctrines strike me as quintessentially "scientific realist" in character, and our eventual response to Poynting's challenge will follow a similar pattern.[5]

Quite generally, nineteenth-century scientific thought reflects the methodological realization that great strides forward in effectively reasoning can be achieved by embedding initial doctrines within richer extended settings that allow (1) for more effective reasoning patterns and (2) perhaps capture external reality better. The unexpected recasting of traditional Euclidean doctrine at the hands of the projective geometers possibly provides the central paradigm of these activities, but enlarged algebraic conceptions of number-like objects were prompted by analogous concerns (Wilson 2020). We have already seen that

---

[4] Descartes labels these notions as "confused ideas," whereas Chapter 5 would consider them as "suitably homogenized."

[5] The central differences lie in the fact that a benevolent deity supplies Descartes with the diagnostic tools required to unravel this engineering upon an a priori basis, whereas we are obliged to work with a ragtag toolkit that we have gradually fashioned for ourselves based upon ancestral inheritance and a large degree of trial-and-error experimentation.

Hertz's own proclivities reflect both of these methodological tenets, there explored, perhaps, to an excessively speculative degree.

Of these two objectives, Richard Dedekind maintained that mathematicians' central motivational duty is to task (1), not (2). In studying a given stretch of reasoning, they should explore its wider neighborhood inside mathematics fully, looking for nearby but improved reasoning opportunities. Doing so often requires a wider overview of a very large catalog of potential reasoning domains. It is in this spirit that Dedekind himself explored the circumstances in which a starting algebra could be extended to a wider frame in which unique factorization is restored. Likewise, mathematicians should investigate what happens when we apply differential equation-like reasoning to complex-valued functions, to stochastic processes, to fractional derivatives, to fractals, etc., even if their motivating interests seem to apply to real-valued domains only (in physics, say). Only within those enlarged settings will we sometimes find the *guideposts* we require to plot our reasoning strategies successfully.

Why is this so? These bootstrapping capacities for exploring wider applicational realms and cobbling together corrected forms of older reasoning patterns represent some of the chief tactics we employ for transcending the parochial limitations on reasoning that we have inherited from our hunter-gatherer ancestors. Even today, we adjudicate the prima facie trustworthiness of a computer program by determining how it performs over a familiar applicational domain, before it is applied to the topics for which it was hopefully designed. Such a check on pragmatic "reliability" is scarcely foolproof, but it is clearly helpful. To be sure, more sophisticated examinations of soundness can supply tighter assurances, but they are frequently hard to construct.

(iii)

Such considerations bring us face to face with another downstream consequence of Carnapian thinking that has seriously confused philosophical thinking on these issues ever since. I have in mind Quine's celebrated criterion of ontological commitment (Quine 1953, 1–19): to determine whether an agent is committed to a range of objects O, force her to articulate her beliefs, as best she can, into the uniform format of a first-order theory T and scrutinize the existential claims she is logically forced to accept, looking for "there is an x such that x is an O." Quine calls this hypothetical process of adopting these theory T proprieties "regimentation" and regards it as a matter of simple intellectual honesty (the following is a paraphrase, not an actual Quine quote):

> If you don't believe in the reality of an ether, witches, or numbers, stop saying things like "there is an ether through which transverse waves travel" and "there is

a witch x that caused a lot of trouble in old Salem." Find some alternative means of expressing the salvable core of what you wish to express within a properly uniformized format. Perhaps these regimented recastings might become "there was a person who was called a 'witch' that occasioned a lot of trouble back in old Salem" and "there are electromagnetic waves which are purely transversal, but there is no ether through which they travel." But the folks who cavalierly claim that "there is an ω-sequence of sets that can serve as a surrogate for the natural numbers" will find it harder to paraphrase away their apparent commitments to a range of Platonic entities. Carnap's contention that such assertions merely "represent 'framework principles'" does not represent an adequately honest answer.

Quine's arguments on this last score appeal to the same gradualist presumptions with respect to linguistic development that we have highlighted. A society may originally presume that "F = ma" represents a framework cornerstone of its physical thinking, but subtle adjustments in profitable modeling technique can dislodge it from that former centrality.

What I find odd about Quine's doctrines lies in their off-handed presumptions that "regimenting" the things we claim in everyday practice represents a comparatively easy task and we can safely assume (as a "convenient philosophical fiction") that it has been successfully accomplished.[6] But this is not so. Our improving standards of practical achievement reflect the inconvenient fact that, in coming to scientific grips with a recalcitrant physical world, we must cobble together the limited collection of sensory and reasoning abilities that we have inherited from our ancestors into novel combinations that allow us to escape from the narrow confines of practicality for which these skills had originally evolved. We have further learned that through exploiting mathematical thinking of a rather purist cast, we can extend and patch together these originally limited resources to more widely varied and reliable purposes. And fresh warnings and refinements with respect to these beacons of reliable reasoning continue to emerge. The early geometers learned that they could strongly certify the correctness of relationships such as the Pythagorean theorem by adding supplementary lines to the figure and by subdividing and moving around its component areas. But the mathematicians of antiquity eventually realized (presumably on the basis of harsh negative experience) that the latter techniques sometimes led to erroneous conclusions (such as all triangles are equilateral). In this fashion were Euclid's tight strictures on constructive proof discovered. But why, exactly, do these two policies of proof so differ in their reliability conferring powers? Kant believed that he had found a philosophical answer, but recent

[6] Better, I think, to recall the old folk song and regard "theories" as "young, but daily growing."

studies in the representational capacities of geometric diagrams have helped address these questions in a more satisfactory manner, in conjunction with a better delineation of our native cognitive resources.

In the same way, we understand (because he wrote about them extensively) the policies of quasi-experimentation that led Oliver Heaviside to his remarkable techniques for extracting extremely valuable and reliable data from linear differential equations. But the underling sources of their deductive surety remained mysterious for many years (in his own assessment, "in reality the logic (of it) is the very last thing" (Heaviside 1912, III 370). In developmental contexts such as these, we must frequently accept as a brute empirical fact that certain veins of mathematicized enlargement and contrast often provide better pathways to reliable conclusion than others, but we may remain quite unsure of what we find ourselves "committed to" in a Quinean vein.

As the central core of this book has argued, our successful multiscalar practices within classical physics likewise resist ready "regimentation" into a single-leveled theory T of the sort that Quine demands. Why these patched together tactics work as well as they do frequently require strategic assessments of a very sophisticated order. There needs to be a "logic" in Heaviside's sense as to why this happens, but teasing it out can prove immensely difficult. But this shouldn't seem surprising. As we previously observed, familiar feats in "mental telepathy" trade upon the art of

Marvelous feats in Mind Reading

extracting enough information from a selected mark X to deductively establish that A holds of X without the innocent victim realizing that she has supplied as much. But the magician who executes the subterfuge can easily learn how to apply the "information sieve" that extracts A from the data provided without himself possessing any strategic understanding of how the "sieve" manages to process the information received (Wilson 2017, 389–91). In such circumstances, a sensible response to Poynting's concerns with respect to "scaffolding removal" demands answers of a considerably different character than Quine encourages.

In the previous chapter, we observed that Theodore Sider regards his prescientific approach to metaphysics as a direct consequence of his self-proclaimed "scientific realism." But this conception of "realism" relies upon the same implausible assumptions of theory T reformulation as Quine does. We'll return to Sider's Panglossian suppositions later in the chapter, after we first review the underlying conception of "scientific realism."

(iv)

Let us first observe that no reasonable conception of "scientific realism" should entangle itself in syntactic requirements of a Quinean character. This chapter begins with a quotation from the physicist Max Planck who is usually regarded as one of the great defenders of scientific realism (he engaged in a famous debate with Ernst Mach on the topic in 1910–11). But he freely acknowledges that "nature does not allow herself to be exhaustively expressed in human thought" (Planck 1932, 1–2), where by "human thought," he means "in mathematized terms." On this understanding, Planck is merely noting the undeniable fact that our successful policies of mathematical registration generally ignore potentially disruptive lower degrees of freedom until such time as explicit attention to their fine-grained complexities is required. As these subdominant disturbances become salient, scientists shift their attention to altered models better suited to these suppressed factors. The only demand that Planck's "realism" places upon these later adjustments appears in the form of the "important and inevitable postulate" that:

> the different hypotheses introduced for dealing with different problems should be compatible with each other. For otherwise the physical picture of the world would lose its uniformity, and in some circumstances we should get two mutually contradictory answers to a definite question.   (Planck 1932, 1–2)

The homogenization barriers characteristic of a standard multiscalar architecture provide an unanticipated resolution to Planck's concerns, for they guard against the emergence of "mutually contradictory answers" without mandating descriptive complexity of an unmanageable character. By working together in cooperative harmony, a suitably monitored patchwork of submodels can capture a wide range of physical behaviors quite admirably without demanding that any single modeling codify all of the target system's varied behaviors within a united and amalgamated narrative. By keeping its "dominant behavior" conclusions suitably stratified, the explosions within descriptive variables characteristic of Tao's "curse of dimensionality" can be avoided.

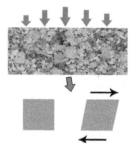

Why model the sheet as a smooth continuum?

Pursuing these considerations a bit further, we can provide a helpful response to Poynting's challenge. As noted in the previous chapter, multiscalar architectures suitable for complex materials often utilize smooth continuum mechanics as the formalism of choice for capturing behaviors upon

Isn't a finite element mesh more realistic?

the higher lengths ΔL. The architecture shifts to a subdivided laminate submodeling only as exceptional events involving lower-scale grain become pertinent to the ΔL-level behaviors. But why should we engage in this continuum physics pretense, given that we are fully aware of the fact that our material is not truly smooth in the manner indicated? Ludwig Boltzmann complains about this artificial descriptive smoothing in his celebrated defense of "atomism":

> The differential equations of mathematico-physical phenomenology are evidently nothing but rules for forming and combining numbers and geometrical concepts, and these in turn are nothing but mental pictures from which appearances can be predicted.... Mathematico-physical phenomenology sometimes combines giving preference to the equations with a certain disdain for atomism.... [D]efinite integrals that represent the solution of the differential equation can in general be calculated only by mechanical quadratures and thus again demand division into a finite number of parts. Do not imagine that by means of the word continuum or the writing down of a differential equation, you have acquired a clear concept of the continuum. On closer scrutiny the differential equation is merely the expression for the fact that one must first imagine a finite number; this is the first prerequisite, only then is the number to grow until its further growth has no further influence. What is the use of concealing the requirement of imagining a large number of individuals now, when at the stage of explaining the differential equation one has used that requirement to define the value expressed by that number?
>
> If I then declare differential equations, or a formula containing definite integrals, to be the most appropriate picture, I surrender to an illusion if I imagine that I have thus banished atomistic conceptions from my mental pictures.
>
> (Boltzmann 1897, 42–3)

What Boltzmann has in mind is this. Consider a sheet of steel, which under a microscope appears as an elaborate warren of crystalline grains. Won't modeling the metal as a numerical array of descriptive units of finite length ΔL remain truer to reality than pretending the sheet is smoothly continuous all of the way down to the infinitesimal-size scale? Shouldn't we accordingly view the numerical "finite element" grid at the bottom of the illustration as more "realistic" than the smoothed-over continuum described in the differential equations that we employ at this stage of our multiscalar modeling?

The methodological naïveté inherent in this passage is astonishing and plainly offers no real defense of atomism at all. But the example returns us to the importance of the *guideposts to modeling reliability* that we discussed in section (iii). Whenever we reasonably can, we should strive to capture a physical behavior within the net of smooth differential equations because the latter represents the mathematical landscape in which we most readily find the clues that warn us that a numerical computation upon a computer may stray from a salutatory path. As we noted in Chapter 5 (iii), it is only at the continuous function level that we can detect the tiny blip that can push a computer simulation into serious inaccuracy. If we remain steadfastly at the level of Boltzmann's "finite division of parts" (i.e., a finite difference scheme of resolution length Δx), we will overlook this warning indicator and deprive ourselves of the helpful signposts of inferential reward and hazard found only

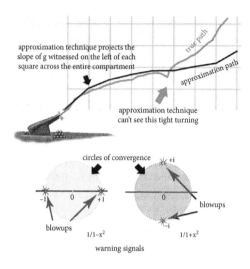

at the infinitesimal level of the differential equations. Similar warnings apply to series summations. Computers don't mind adding up whatever factors we give them, no matter whether the outcomes are descriptively nonsensical or not. But the standard indicators of summation failure often lie in the singularities located upon the complex plane outside of the real line (Wilson 2017, 393–4). Indeed, a computer can't distinguish a true infinitary blowup from a really big number unless we somehow supply it with guiding criteria. But the "stop" and "go" signals required to monitor the trustworthiness of a simple series summation commonly require expeditions into further realms of faraway structure, sometimes of a quite exotic stripe.[7]

A subtler version of this same misunderstanding is found in Poston and Stewart (1978, 300), with respect to the subtle functional analysis considerations one now finds in a sophisticated engineering primer:

> We have gone beyond the casual assumption of "configurations pretty much like smooth functions if we don't look too closely".... We are making very delicate hypotheses (like square integrable second distributional derivatives) about exactly what we would see if we could look arbitrarily close: hypotheses we know to be false. In what sense these "infinitesimal" hypotheses can be said to model the local average nature of elastic solids is a dark and gleaming mystery. The results of the model, in macroscopic predictions, are very successful indeed; but it is hard to say in what sense the theory is more exact than a computer

---

[7] From this point of view, blow-up singularities on the complex plane frequently supply us with commonly represent our most effective instruments of physical data registration. See Wilson (2006) for more on this theme.

treatment of a finite element model like that of section 4, but with many more rods and strings. ... Long calculations that you do not understand, whether by number crunching or Banach space methods, will land you with bridges that fall down. A theory is only as good as your understanding of it.

I'll supply a fuller rejoinder below, but the crucial guidance required to construct a reliable computer program for estimating the internal stress within a notched metal plate directly depends upon the delicate features of the "square integrable second distributional derivatives" of which Poston and Stewart complain. Once again we must peek into a rather daunting corner of Greater Mathematicsland to find the practical guidance we require.

In a similar vein, I have sometimes heard critics wonder, "Why should we fuss so much about whether the Navier-Stokes equations (which represent a severely smoothed over continuum account of fluid behavior) genuinely support a transition to turbulence? Water, after all, is decidedly molecular in composition." A rigorous positive answer (which is generally regarded as one of the greatest outstanding problems within applied mathematics) will provide us with an abundance of useful information with respect to questions as to whether turbulence can be usefully captured, say, in terms of the speculative energy cascades posited in Chapter 6. Our abilities to successfully capture higher-scale dominant behaviors in manageable mathematical terms strongly dictates how we might cobble together a descriptive architecture responsive to its richer complexities.

Attention to these naturalistic considerations strikes me as essential to a just appraisal of what I like to call "our computational place in nature," viz., in view of the physical environments in which we live and the basic computational skills with which we find ourselves supplied, which descriptive tasks will we find ourselves able to execute and which not? And which foreseeable  avenues of improvement still remain open to us? Quasi-biological considerations of this type comprise the primary topics of Wilson (2017, 30–2).

Pace Kantian contentions otherwise, these observations with respect to the palpable imbalances that characterize human inferential capacity may simply reflect inherited limitations upon our biological capacities for ampliative reasoning, similar to the factors that inhibit a lungfish's navigational capacities when it first clambers upon the land (I believe that Mach would have largely concurred in this opinion). But, like that intrepid ichthyological adventurer, we can gradually improve our practical capacities. By employing clever mix-or-match reasoning tactics, we can patiently work our way past many of our initial limitations in reasoning, leading to an improved and nuanced appraisal of our computational circumstances, along both favorable and unfavorable directions (we now recognize that we can successfully bisect an angle with compass and ruler, but not obtain a trisection, but our hunter-gatherer ancestors could not have grasped this fact so firmly). Perhaps the angels needn't reason in this circumscribed manner, but we must. Those empyrean entities haven't arisen as the raw products of natural evolution, but we have, and we must live with the limited abilities we've received.

None of this indicates that the availability of reliable numerical methods (such as a finite element computation) isn't important to multiscalar modeling as well. Indeed, it is only through such means that we will normally be able to estimate localized stresses within a loaded metal sheet of a minimally complicated geometry. It is merely that the analytic clues found within their motivating set of differential equations are needed to adjudicate the soundness of the numerical scheme itself (which is notoriously capable of cheerily computing completely spurious results if left unmonitored).

(v)

These last considerations point to an important issue that becomes utterly muddled through Quine's insistence upon amalgamated regimentation. These confusions arise from the fact that advancing large swatches of mathematics effectively, our perceived interests in their "references" typically shift to a significantly different basin of concern. Richard Dedekind explicates this divergence as follows:

> If one finds the chief goal of each science to be the endeavor to fathom the *truth*—that is, the truth which is either wholly external to us, or which, if it is related to us, is not our arbitrary creation, but a necessity independent of our activity—then one declares the final results, the final goal (to which one can in any case usually only approach) to be invariable, to be unchangeable. In contrast, science itself, which represents the course of human knowledge up to these results, is infinitely manifold, is capable of infinitely different representations. This is because, as the work of man, science is subject to his arbitrariness and to all the imperfections of his mental powers.... The introduction of such a [computational] concept as a *motif* for the arrangement of the system is, as it were, an hypothesis which one puts to the inner nature of the science; only in further development does the science answer; the greater or lesser *effectiveness* of such a concept determines its worth or worthlessness.   (Dedekind 1854, 755–6)

As we've just seen, the most effective signposts to reasoning improvement and correction frequently lie outside the mathematical arena in which one originally begins (within physics, for example). Dedekind emphasizes the importance of the wide-ranging examination of reasoning "motifs" across a vast array of coherent settings, whether or not they directly bear upon any physical application or not. Quite frequently, these novel domains are first reached through some form of appeal to what Dedekind calls "persistence of theorems," the recognition that superior reasoning policies can be obtained if the irregularities that block deduction within previously conceived arenas can be removed through supplementary "ideal" ingredients.[8] Nevertheless, the terminologies that we retain in pursuing these exploratory inquiries can sometimes prove problematic, for subtle difficulties sometimes emerge when linguistic expressions that superficially appeared trustworthy fail to find adequate supports within the newly extended setting (the simplest example of such a referential failure is "6/0," but Riemann's unchecked appeals to "minimal surfaces" singlehandedly generated the greatest single crisis in nineteenth-century mathematics: the failure of the Dirichlet Principle) (Monna 1975). As a result, a mathematician will understandably appropriate the managerial utilities of "refers" to her own investigative purposes, despite the fact that any significant concern with their direct real-world support will have been left far behind. It is exactly in this context that Richard Courant highlights the parochial importance of "mathematical existence":

> But empirical evidence can never establish mathematical existence—nor can the mathematician's demand for existence be dismissed by the physicist as useless rigor. (Only a mathematical existence proof can ensure that the mathematical description of a physical phenomenon is meaningful.)   (Courant 1950, 3)

However, as our attempts to adjudicate the merits of our deductive practices within the "mixed modalities" characteristic of applied mathematics research, both claimants on the notion of "reference" can naturally come into play (e.g., "Jan Mikusiński ably established mathematical referents for the inverse operations within the Heaviside calculus, but it's dubious that he thereby captured their physical referents in an entirely happy way. Schwartz's distributions are closer to the mark in those regards").

Quine's insistence upon unified regimentation immediately blurs these distinct purposes with respect to "existence" into an unhelpful muddle. Penelope Maddy has criticized these significant shifts in focal concern with respect to "reference" in a detailed manner, although sometimes in terminologies that I regard as suboptimal (Maddy 2007; Maddy 2020), (I prefer Dedekind's emphasis upon mathematics' interest in exploring various "motifs for the arrangement of a system")

---

[8] Two of Dedekind's motivational paradigms are the projective revolution within Euclidean geometry and Kummer's introduction of "ideal factors" into algebraic number theory. He then attempts to build up the familiar number systems utilizing analogous tactics (Wilson 2020).

(Dedekind 1854, 756).[9] From this point of view, set theory gains its importance not by providing an all-embracing "ontology" for mathematics but as a set of extremely valuable tools for investigating the internal consistency of various mathematical "motifs" and their potential embeddings within one another (Wilson 2020).

These observations deepen our appreciation of why brute "constant conjunction" approaches to causation fail to capture the superior appeal of the smoother differential equation treatments that du Châtelet favors, albeit in a less metaphysically laden manner. We can still applaud her perceptive emphasis upon the importance of smooth differential equation behavior, even when some of this "importance" stems from the surprising manner in which the infinitesimal realm in which these smooth curves mathematically dwell provides the guideposts that humans require to reason reliably about the dominant behaviors of higher scale ensembles.

(vi)

These developmental considerations suggest that a suitable "realist" reply to Poynting's concerns with respect to "scaffolding" can be provided in a "reverse engineering" mode much like Descartes's. If we consider a multiscalar architecture suitable to a piece of steel, we can fairly easily distinguish the "guideposts to effective reasoning" aspects of the scheme from the natural behavioral opportunities to which they respond. Such divisions are common within any form of biomechanical explanation: we can distinguish the behavioral capacities of a colony of honeybees from the opportunities inherent in the pollen distributions that they seek to exploit. The fact that the shearing behavior of a moderately sized steel specimen is governed on a large scale by two bulk parameters and is able to largely regain its original shape when the applied force is removed represents a descriptive opportunity that resides within the metal itself. In contrast, the fact that we reason most effectively about these behaviors using smooth

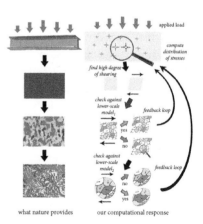

what nature provides   our computational response

---

[9] In particular, the terminologies that Maddy employs to this end ("arealism" and "thin realism") strike me as unhappily suggestive of a deflationist literature with which I have little sympathy. She wishes to distance her views from these contentions as well (Maddy 2022), but I worry that unwanted connotations linger on in her formulations. Possibly I'm overly sensitive on these scores, for standard deflationists trivialize the very investigations into word-to-world relationships that I regard as central to a "scientific realism" properly conceived. Semantic minimization of this character is utter anathema to me.

continuum physics models significantly reflects the internal structure of our human reasoning capacities, because such a placement supplies the largest array of clues to help us calculate reliably and to understand and improve the strategic recipe underlying those calculations. We should accordingly respond to a Boltzmannian critic as follows: Of course, we realize that the steel is composed of molecules, but it's better to not employ that kind of language with respect to its higher-scale behaviors, because we want to focus as best we can upon the mathematical ingredients that most effectively align with the behavioral opportunities available at the scale length of large populations of such molecules. When their finer-grained behaviors become relevant, we will shift to replacement sub-models better suited to such details, but we should only wish to make those modeling adjustments as circumstances require. Nothing within this sensible "division of linguistic labor" strategy suggests that we don't believe in molecules, but it does explain why we don't explicitly talk about them all of the time. In a similar manner, we can distinguish nature's environmental offerings (how water behaves) from the reasoning policies that promise the greatest rewards of tractability and security in response. At present, the Navier-Stokes equations appear to be the best we can offer in the latter category, but their intrinsically mathematical behaviors remain so uncharted that it's hard to be sure.

In other writings (Wilson 2006), I've identified various ways in which our everyday employment of words like "color" and "hardness" divide into local patchworks that interconnect with one another in allied ways. Any reasonable answer to the philosophical chestnut "to what physical property does the word 'red' correspond?" can reasonably be addressed only if we consider the varying platforms of classificatory opportunity provided within spectral reflection in conjunction with the discriminative capacities of the human eye. And I write "varying platforms" because the relevant informational opportunities frequently vary considerably from material to material and environmental setting to environmental setting. Thus the directional aspects of a metallic object's shiny surface supply considerably richer information about its surface electronic structure than a plank of wood coated with a flat paint (we shall revisit these considerations in Chapter 9 when we consider the utilities of the word "reference"). An adequate unraveling of the complexly entangled "architecture" that governs our everyday "color" discrimination would comprise a significant achievement in "reverse engineering." But this is the patient manner[10] in which I believe that Poynting's challenge should be addressed. We should study how the determinants of practical need and sensory capacity progressively factor a usage into distinct but mutually

---

[10] Pincock (2010, 120) has characterized my point of view as "patient realism." With respect to color vocabulary, I became so overwhelmed with their complexities that I gave up trying after a point in Wilson (2006). I became subsequently enamored of the admirably developed architectures of modern multiscalar modeling precisely because they effectively illustrate in a controlled manner many of the linguistic developments that I had originally hoped to highlight.

intercommunicating sectors, whose supportive rationales commonly rest upon significantly different forms of physical underpinning.

In many ways, mathematical and philosophical thought have passed like trains in the night along opposing tracks ("a-whooee, a-whooee"). Both lines of thought initiate within the simple ordinary differential equation (ODE) conception of theoretical content exemplified within the point mass approach to mechanics outlined in Chapter 3. It was within this sheltered mathematical setting that the celebrated French atomist Pierre-Simon Laplace formulated his celebrated articulation of predictive determinism:

> We may regard the present state of the universe as the effect of its past and the cause of its future. An intellect which at a certain moment would know all forces that set nature in motion, and all positions of all items of which nature is composed, if this intellect were also vast enough to submit these data to analysis, it would embrace in a single formula the movements of the greatest bodies of the universe and those of the tiniest atom; for such an intellect nothing would be uncertain and the future just like the past would be present before its eyes.
>
> (Laplace 1814, 4)

In these restricted circumstances, many of the complexities that generate the doctrinal conflicts I have emphasized have not yet clearly emerged. We have not yet witnessed the fruitful but problematic amalgamation of force-based ODEs (**F** = m**a**) with constraint equations that lies at the heart of the tensions that trouble Hertz.[11] Nor have the rich and subtle questions of descriptive harmony that bedevil standard appeals to "boundary conditions" appeared, because "boundary conditions" properly characterized generally appear after applied mathematics shifts into the territory of the partial differential equations (PDE) that are utilized in higher-dimension continuum physics.

Here's a vivid illustration of how these boundary condition complications typically emerge. Suppose we model a metal plate with a suitable set of interior equations (Laplace's equation for a membrane will suffice) and assign it suitable boundary conditions by pushing or pulling on its outer boundary in normal ways. We then confront an escalating set of tensions between interiors and boundaries in the following way. (i) if the boundary is a smooth curve such as an ellipse, interior and boundary demands will accord with one another in the unproblematic matchups that one studies in an introductory class. (ii) Insert some convex corners in the plate, and the previous boundary/interior accord will break down in those locales. But the strain energy within any finite sector will remain finite, so we

---

[11] More exactly, these combinations were widely utilized, but the complications that constraints introduce with respect to the intuitive existence and uniqueness features of purist ODE settings were not recognized until later in the century.

can reformulate
our mathematical
expectations as to
how "solutions"
should meet their
"boundaries" in a
manner that can
still be regarded as

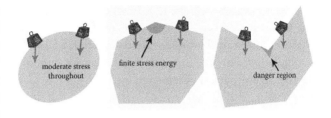

a suitable account of the energetic situation within the plate. (iii) However, introduce some inset corners into the boundary, and any attempt to supply a viable "solution" will contain patches of strain that blow up to impossible infinities here and there. No real material can sustain strains of that nature. Nonetheless, these same "impossible regions" remain descriptively useful, for real-life cracks generally initiate in these same stress concentrating corners. As such, a liberalized notion of "solution" that tolerates internal infinities can still provide engineers with the warning signposts they require to investigate those particular regions more closely, by consulting a revised modeling that introduces tiny cracks into the strained region's previously unblemished interior (we then determine whether sufficient driving energy is available to elongate these cracks). But to render these determinations mathematically coherent, we must twiddle once again with our notions of "solution" and "boundary accord" (doing so generally demands a fair amount of functional analysis sophistication).

In fact, these are the exact considerations that explain why the subtleties of the "square integrable second distributional derivatives" of which Poston and Stewart complain become so important to the practical engineer. These provide the critical *warning signals* she will need in simulating the behavior of a metal plate upon her computer in a trustworthy manner. This need to search for relevant consider-ations within a wider stretch of Greater Mathematicsland supplies a further exemplar of the response we offered earlier to Boltzmann.

The conceptual ramifications of these shifting requirements upon interior/ boundary accord are nicely captured within a pungent observation from the mathematician Hans Lewy:

> [Mathematical analysis only supplies] kind of hesitant statements.... In some
> ways analysis is more like life than certain other parts of mathematics.... There
> are some fields ... where the statements are very clear, the hypotheses are clear,
> the conclusions are clear and the whole drift of the subject is also quite clear. But
> not in analysis. To present analysis in this form does violence to the subject, in
> my opinion. The conclusions are or should be considered temporary, also the
> conclusions, sort of temporary. As soon as you try to lay down exact conditions,
> you artificially restrict the subject.   (Reid 1991, 264)

These complications comprise some of the mathematical reasons why we should not frame our theory T conceptions of "how science works" according to the simple contours of Newtonian celestial mechanics conducted within a point mass setting (where only the simplest ODE tools are required). As P. G. Tait notes in a passage already cited in Chapter 5:

> [T]he simplicity of the data of the mathematical problem is gone; and physical astronomy, except in its grander outlines, becomes as much confused as any other branch of science.   (Tait 1895, 9–10)

We can no longer luxuriate in the unsullied sunshine of the Laplacean mind, but I believe that theory T thinking does exactly that.

However, I have begun to dip deeper into the pool of mathematical technicalities than the central morals of this book require. My guiding intention is merely that of highlighting the characteristic manner in which the tiny confusions and misalignments that everywhere accompany effective description represent subtle tokens of some gradually evolved, but often unnoticed, backgrounded architecture. It is through tacit reliance upon these unnoticed controls that human reasoners manage to extend the applicational ranges of our discourses and correct for their deficiencies. But if these observations are well founded, then the last thing that a philosophical inquirer should wish to do is to smother this vital symptomology beneath an ill-conceived layer of theory T "regimentation."

(vii)

Our chief reason for concentrating upon physics examples in this book lies in the fact that the subject's parochial concerns have served, historically, as the epicenter from which inappropriate conceptions of "rigorous" inquiry have gradually infected other parts of the philosophical corpus. The conceptual clarity that Hertz's axiomatic proposals once promised to the classical physicist continues to serve as the ersatz standard of "rigor" to which a proper "philosophical theory" should piously strive. The fact that genuine conceptual improvements within mechanics have demanded wider-ranging and far messier forms of corrective investigation has mattered little, because the philosophers stopped attending to the original motivating problems. Insofar as I can determine, the wrong turning centrally responsible for coronating this inadequate conception of the diagnostic enterprise as philosophical royalty lies in the mid-twentieth-century manner in which Hertzian axiomatics become rarified into the vagaries of theory T thinking. In so doing, Nagel-like appeals to logic-focused notions such as his versions of "lawlike generalizations" and "initial and boundary conditions" replace the original discriminations actually required within the classifications that remain mathematically accurate. Hertz himself is not to blame for these later abuses; his own diagnostic work was truly exemplary. It is merely that he suggested the wrong solution to the tensions that he had correctly noticed.

This deficient circle of logic-like classifications has persuaded most of us that we can happily embrace a schematic vision of "theory" that ignores the very details to which philosophers should most carefully attend, in whatever field they hope to scrutinize. In the previous chapter, we reviewed the manner in which analytic metaphysicians commonly justify their labors as attempts to formulate science-like "theories" of their chosen topics of interest. But this concept of "theory" is entirely mythological in origin, tracing to axiomatic hopes that never material-ized).[12] It is within this pseudoscientific vein that the followers of David Lewis justify their speculations. Quoting Theodore Sider once again:

> [M]etaphysicians use standards for choosing theories that are like the standards used by scientists (simplicity, comprehensiveness, elegance, and so on).
>
> (Sider 2007, 6)

It is for these basic reasons that Carnap's "fiction of a completed construction of physics" cannot be tolerated as a "harmless epistemological convenience" but serves as an extirpative presumption that erases the delicate detail required to identify and defang the conceptual obstructions that impede developmental progress. On the basis of a Pollyannish conception of "scientific realism," writers like Sider have concluded that a "completed construction of physics" free of these complex forms of submodeling entanglement will inevitably become available, ready to supply metaphysicians with an uncompromised inventory of the world's "fundamental kinds." With one swift snicker-snack of the vorpal blade, he dispatches all of the troublesome details we have highlighted. But this same tactic vacuums up all of the vital symptomology required to unravel any of the robust mysteries that emerge as the natural collateral damage of any naturally occurring conceptual advance.

For these reasons, we should not transfer Hilbert's dictums on "rigor" uncrit-ically to applied mathematical practice:

> Indeed, the requirement of rigor, which has become proverbial in mathematics, corresponds to a universal philosophical necessity of our understanding; and, on the other hand, only by satisfying this requirement do the thought content and the suggestiveness of the problem attain their full effect.   (Hilbert 1902, 441)

---

[12] This same folklore particularly suits the armchair proclivities characteristic of the "analytic" enterprise. In former times, the term "armchair philosophizing" was regarded as pejorative, but the modern metaphysics school appears to have adopted this designation as a badge of honor. I once attended a conference that proudly featured a comfy chair upon its advertisement poster (and the event was housed in a building dedicated to Emmy Noether, no less).

This alleged "necessity of our understanding" is cut from the same philosophical fabric as the traditional assumption that the rules of sound methodology are governed by a "universal causal principle" that demands that we must capture nature's behaviors within a fabric of ill-considered "laws of nature." This latter requirement is both vague and patently excessive, and largely stems from the inferential advantages of employing partial differential equations within a wider fabric of corrective feedback modelings. Conducting our researches with full Hilbertian rigor is generally advisable when we pursue goals of a "purely mathematical" character, but these same advantages significantly dissipate when we enter the "indefinite" arena of partial differential equations, where we must frequently decide how to patch sundry shards of mathematized reasoning together in a manner that benefits our understanding of nature. This kind of assembly work may demand a "rigorous" attention to subtle detail, but it is merely the informal rigor of careful consideration that is wanted, not conformity to some dubious standard of preset "universal methodological necessity." As Heaviside puts the point, our first order of business is "to find out what there is to find out" (Heaviside 1912, II 2).

In a nutshell, these observations capture the complaints with respect to "ersatz rigor" with which this monograph began. Internally closed studies of a Hilbertian stripe are frequently quite valuable, but they fail to address a significant layer of further concern with respect to how the individual tools of mathematized calculation should be cobbled together in a manner that befits our hopes of effectively capturing real world behaviors within workable webs of mathematized netting. Occasionally a philosopher might establish a worthy theorem of a Hilbertian character, but these should not be viewed as the primary goals of our profession. The skills we should instead cultivate are those derived from a rich appreciation of how problematic tensions infect a wide variety of otherwise practical forms of discourse, coupled with a thorough acquaintance with the subtle symptomologies of concealed backgrounded architectures. Nothing widely universal can be promised upon a methodological front, because the rewards and punishments of evolutionary development frequently carry their captives to the most astonishing and unexpected locales. But appreciating these widely ranging varieties of remedy can prove useful in itself, for one can feel more secure in "thinking outside the box" if consulting experts are available who have cultivated a cross-disciplinary appreciation of how similar obstacles have been previously surmounted within an array of fields. These consulting skills, I propose, provide a better model for the philosophical enterprise than the attempts to fashion ersatz rigor "theories" based upon armchair experience. This book's chief complaint with respect to contemporary practice is that its currently admired standards have allowed these informal diagnostic skills to shrivel to virtual minimality, rather than comprising the vein of productive suggestion from which an academic philosopher might obtain the rewards of justified satisfaction.

For the reasons indicated above, I have expressed this book's intended lessons in largely philosophy of science terms. But I firmly anticipate that similar observations apply to other topics of a different stripe. Indeed, J. L. Austin's stated rationale for investigating "excuses" within the context of moral evaluation can be construed as a close copy, mutatis mutandis, of the managerial tasks that interconnected mathematized patchworks perform in sustaining the comparable "mechanisms" of a multiscalar enterprise. He writes:

[T]o examine excuses is to examine cases where there has been some abnormality or failure: and as so often, the abnormal will throw light on the normal, will help us to penetrate the blinding veil of ease and obviousness that hides the mechanisms of the natural successful act. It rapidly becomes plain that the breakdowns signalized by the various excuses are of radically different kinds, affecting different parts or stages of the machinery, which the excuses consequently pick out and sort out for us.   (Austin 1966, 127–8)

In any case, with respect to Sider's appeals to "realism" at the end of the previous chapter, no reasonable formulation of "scientific realism" should demand that future science will suit the schematic dictates of Sider's armchair metaphysics. Here's an analogy to capture the descriptive circumstances we may confront. In the movie *Forbidden Planet*, the fearsome Monster from the Id is invisible, but the crewmen from Starship C-57D can still mark out its positions by refracting tracer bullets along its contours. The Monster is undoubtedly real (it crushes Lt. Farman, inter alia); it is

merely that its outline within the tracer contrails is fragmented. And only in this gradualist manner will descriptive inquiries conducted in mathematized terms find themselves able to corral empirical reality. As Planck says, "nature does not allow herself to be exhaustively expressed in human thought" (Planck 1932, 2). Nonetheless, we can improve our descriptive lot considerably if we cleverly link together patches of mathematical technique informed by a wider appreciation of viable recipe.

On the other hand, we must also learn when to abandon a cherished descriptive task when nature churlishly refuses to cooperate with our desires. On these scores, advanced mathematics has reached many negative assessments with respect to inferential tasks that we might wish to accomplish but can't do so feasibly. We might vainly hope to square the circle or construct a perpetual motion device. Or establish the stability of the solar system. Or enumerate the highly non-recursive succession of numbered posts that a careening ball might progressively strike within a frictionless pinball machine. Nature will not cooperate with any of these ambitions, and sophisticated mathematical tactics can often tell us why.

(viii)

These negative obstacles represent significant checks upon the more sanguine forms of descriptive optimism that Hertz's axiomatic admirers have unwisely embraced. In Chapter 4, we quoted Ernst Cassirer's unbounded admiration for Hertz's holistic approach to conceptual analysis (leading to a situation where "our minds, no longer vexed, will cease to ask illegitimate questions" (Hertz 1894, 8)). Cassirer concurs in Hertz's recommendation that proper conceptual repair should investigate the practical advantages that a troublesome word like "force" supplies within the full applicational nexus to which it belongs, rather than vainly attempting to isolate some internal rottenness specific to the concept itself. Approaching individual "meaning" in the traditional vein of word-based internal "conceptual analysis" may only unleash the genies of ancestral prejudice that frequently serve as impediments to scientific advance (as Mach frequently emphasizes, earlier stages in word usage never completely die away; they merely become latent, ready to be reinvigorated when the bottle is opened). These are the reasons why Cassirer frequently compares a conceptual arrangement to an integrated "machine" or "organism." Quoting him once again:

> Physical science is a system that must be accepted as a self-contained whole, an organism of which no single part can be made to work without all the others, even those farthest removed, becoming active—some more, some less, but all in some precise degree.   (Cassirer 1950, 113)

But a flat descriptive landscape of the type Cassirer seemingly envisions does not necessarily constitute the scientist's *summum bonum*, for the reasons that Tao's "curse of dimensionality" vividly illustrates. We can obtain an excellent appreciation of everything that transpires within a hefty hunk of granite by utilizing a wisely devised multiscalar scheme, despite the fact that the narratives that we spin out in this manner will not unfold smoothly but instead engage in various forms of footnote-justified adjustment pointing in sundry directions (resulting in a narrative closer to *Tristram Shandy* than *McTeague*). Proceeding otherwise may prevent us from seeing the rock but for the grain. The demerits of such an ill-focused enterprise are warmly acknowledged by many of the "realist" scientists we have cited (viz., Planck, Thomson, and Tait).

As a consequence, the task of an effective reformer is often that of correcting for the inadvertent errors of an established practice while leaving its most vibrant inferential landmarks more or less intact. The result may wind up like an ancient city that has been refitted for modern usage through an elaborate warren of underground tunnels that allow the information stored in building A to flow freely to building B despite the fact that such a communication channel was not envisioned when the city was originally constructed. Throughout this book we have cited the oddly gerrymandered "operational calculus" that the English autodidactic Oliver Heaviside developed in response to the mathematical perplexities of electrical circuits (Wilson 2006, 529–45; Mahon 2017). Through trial-and-error experimentation, Heaviside removed the obstructive barriers within the conventional treatment of differential equations by supplementary "extension elements" (e.g., introducing inverse operations where none had existed before). He cheerfully confessed that he didn't understand the information-processing maneuvers that allow his peculiar "calculus" to reach correct conclusions with their striking efficiencies:

> But then the rigorous logic of the matter is not plain! Well, what of that? Shall I refuse my dinner because I do not fully understand the processes of digestion? No, not if I am satisfied with the result.... First, get on, in any way possible, and let the logic be left for later work.   (Heaviside 2003, III 9, 461)

But such "later work" is clearly wanted, because Heaviside's modified rules are not completely infallible and rely too heavily upon the "physical intuitions" of the calculator. The task of figuring out from an informational point of view what transpires beneath Heaviside's strange manipulations was bequeathed to a later generation of applied mathematicians. They eventually determined that mathematical formulas that had been formerly parsed as δ/ε point-focused differential equations had shifted to reflect integral equation relationships involving smoothing operations spread over smallish volumes. And they further realized that their reanalysis supplies a better account of the word-to-world relationships that had prompted the use of the original modeling formulas in the first place (Wilson 2017, 324–61).

A similar pattern of revisionary thinking has affected modern classical mechanics practice. The availability of adequate computers originally facilitated the policy of linking submodels together into architectural arrays. The successes obtained by automating intuitive checks-and-balances interconnections in this operational manner subsequently prompted applied mathematicians to subject the rough homogenization policies they employed to a higher degree of rigorous scrutiny.

In these contexts, Heaviside's "logic" signifies a subsequent diagnosis of the underlying "information processing maneuvers" that must roughly underpin the

majority of inferential transitions permitted within his calculus without necessarily agreeing with all of them. Quite commonly this underlying "logic" proves to be
quite different than what the calculus's original employers had once originally
fancied (Wilson (2006) describes this as a faulty "semantic picture" of the usage's
machinery). I have several times cited a close analogy in our earlier discussions:
the card-guessing routines that are easy to implement but whose information-
processing strategies demand substantial mathematical excavation. To protect
against applicational error, we generally want to study how information is passed
along the reasoning routines permitted within the novel calculus, in part to gain a
sharper appreciation of where such routines are likely to lead to error. Computer
scientists dub these informational tracings as "verifications of correctness,"
whereas logicians label them as "soundness proofs." These represent the after-
the-fact investigations we will wish to pursue after we have roughed out the
network of inferential connections at play within a practical usage. We will want
to understand the objective landscape that secretly informs the successes of an
evolved calculus such as Heaviside's, without ratifying every claim that it certifies.
To repair its faults, we should endeavor "to follow the information" and repair the
unavoidable glitches that will have inevitably crept in. In doing so, a variety of
somewhat different reconstructions may prove viable (Heaviside's calculus has
been "rigorized" in a variety of ways). Such variations with respect to the
improvement of an inductively established reasoning recipe are to be expected
and reflect the manners in which such practices have gradually solidified from a
more amorphous set of practical skills. Biological evolution rarely produces
perfectly designed specimens, and the same moral applies to linguistic development as well.

Wilson (2006) dubs these successive stages in linguistic reform as "seasonalities" whose characteristic phenomenologies we shall explore a bit further in
Chapter 9. The central moral with respect to our present concerns resides in the
observation that it is exactly along these lines of "seasonal" exploration that a
scientific realist should expect to uncover the considerations that distinguish the
"scaffolding" aspects of a successful descriptive practice from the objective opportunities that support it.

Sider's version of "realism" does not proceed in this manner at all, but instead
awaits the hypothetical "fundamental physics" that Sider assures us is someday
forthcoming.[13] From its nicely regimented contours he believes that an eager
metaphysician will be able to extract its "ontological commitments" in Quine's
recommended manner, bypassing all of the messy complications upon which
I insist. How have Sider's utopian expectations managed to plant such deep

---

[13] Even for "theories" like classical mechanics that eventually lose the guiding imprimatur of nature.
Sider appears willing to appeal to "the possible models of classical mechanics" in the absence of any
evident set of axioms that might successfully carve out such structures.

roots within our academic psyches? How has a proposal for methodological reform that didn't work (Hertz's axiomatization) turned into a "convenient idealization" that philosophers can confidently embrace within their methodological musings? Why has philosophy of science largely abandoned the careful distinctions of the applied mathematicians in favor of the wan schematisms and ill-defined "laws of nature" characteristic of theory T thinking?

(ix)

Let us now review several further incidents that have contributed to this strange transmogrification. Recall Hertz's and Helmholtz's unfortunate contrast between the "gay conceptual garments" that enfold the mental conceptions (Hertz's "images") that we vibrantly grasp and the dimmer exterior properties that we can only consider in the skeletal manner of a Harpo-imitates-Groucho correspondence.[14] Let's call the vocabularies whereby we reach out to these minimally grasped properties-in-themselves as "theoretical vocabulary" V_T. Influenced by Hertz and Hilbert,[15] philosophers such as Moritz Schlick conceived the radical ambition that an adequately closed axiomatic system A might completely settle the truth-values of sentences that contain these "theoretical" V_T terminologies through "implicit definition": as soon as the extensions of the observational predicates within T become determined, A's axiomatic demands will force unique truth-values upon the V_T remainder. If so, we needn't worry further about word-to-world underpinnings of such words because their contributions to sentential truth-values can be estimated from A-internal considerations alone. Such hopes frame the motivational basis for the brash manifesto cited earlier:

> Philosophy is to be replaced by the logic of science—that is to say, by the logical analysis of the concepts and sentences of the sciences, for the logic of science is nothing other than the logical syntax of the language of science.
>
> (Carnap 1934, xiv)

Fulfilling this program encountered many familiar snags in the years to follow, but hazy doctrines of an "implicit definability" character live on to this day, including Thomas Kuhn's notorious contention (Kuhn 1962) that paradigm-dependent "conceptual incommensurabilities" prevent successive scientific

---

[14] Worrall (1989) strikes me as cut from this same conceptual cloth.
[15] Hilbert (1925) hopes that standard objections to the extension elements within geometry and infinitary quantifications within logic might be painlessly avoided in this hermetically enclosed syntactical manner.

generations from comprehending one another.[16] Shortly thereafter, Hilary Putnam hoisted a banner of self-proclaimed "scientific realism" in which he reasonably objected to Kuhn's extreme position on the grounds that scientific dialectics actually concern a common referential basis upon which all parties actually agree, whether the "theories" they endorse are similar or not (Putnam 1975). In particular, Putnam contended that the disputants in controversies concerning, e.g., "temperature" were all concerned with the focal characteristic of "possessing an averaged molecular kinetic energy of fixed value E" without realizing that they were doing so. Based upon this de facto identification of referential target, we can meaningfully assess which of the parties had advanced correct claims with respect to "temperature" and which had not.

Let us here note that historians of science commonly employ allied referential idioms of "reference" in identifying a central objective focus within a scientific practice. In this vein, we can usefully declare that "When Thomas Young spoke of 'monochromatic light,' he was actually *referring* to a power spectrum decomposition of a statistical steady state ensemble stretching over a sizable temporal interval encompassing a large number of pulse-sourced events." In doing so we can simultaneously acknowledge the important lineage of optical research that Young initiated as well as the surprising revisions in our conception of "spectrum" that became later recognized. As the mathematician Robert Woodhouse once wrote, "a method which leads to true results must have its logic":

> Now if operations with any characters or signs lead to just conclusions, such operations must be true by virtue of some principle or other; and the objections against the [facilitating quantities] ought to be diverted upon the unsatisfactory explanation given of their nature and uses.   (Woodhouse 1802, 90)[17]

This is the same sense of "logic" that we have encountered in Heaviside: an unraveling of the mysterious strategic patterns that explicate how input information becomes secretly processed to effective ends within a successful descriptive practice. As we have already noted, practitioners at sundry developmental stages within a usage's evolving career frequently entertain wildly inaccurate conceptions of the informational processes at issue, in the manner of an amateur magician who believes that he can really read minds. Sympathetically evaluating the typical motley of partially just and partially unjust conclusions that characterize the historical record often proves context sensitive in extremely irregular ways.

---

[16] Kuhn's exact reasons remain rather inscrutable due to his reliance upon the murky contours of his "paradigms." Other authors have attributed allied "incommensurabilities" to framework disagreements conceptualized in Carnap's general manner.

[17] I first learned of this passage from Smith (1851). Woodhouse's original essay is more conservatively intended than Smith's citation suggests, which isn't surprising in view of the radical reshapings that mathematics underwent in the half century that separates these authors.

Nonetheless, identifying their proximate objective supports can prove enormously enlightening in appreciating the unexpected twists and turns that our evolving descriptive policies evidence as they gradually wend their ways to improved performance. I believe that these adjustments represented Putnam's original focal concern. I was a thesis student of his and was greatly inspired by the suggestion that the history of science might serve as an excellent laboratory for directly studying the ways in which concrete word-to-world relationships progressively adjust and refine themselves under the goad of practical achievement. Because the resulting word-to-world discriminations commonly prove very complicated, some parties have suggested that the non-trivial evaluative uses of "reference" should be jettisoned altogether (Field 2001). I thoroughly reject such defeatism. Yes, the byways of language are often byzantine, but that's simply an important aspect of the natural history of *homo sapiens* with which we must contend in any useful form of "philosophizing." I shall append a few supplementary remarks on the value of diagnosing non-trivial word-to-world relationships in terms of a semantically liberated notion of "reference" in our final chapter.

For reasons I never fully understood, Putnam abandoned these straightforward "study conceptual development within the history of science" policies in favor of a rigid form of alleged "scientific realism" that he articulared (in company with Richard Boyd) as a developmental slogan "The terms in a mature science typically refer" (Putnam 1978, 21). This thesis anticipates that the development of any worthy body of scientific doctrine will eventually reach a condition in which its vocabularies will subscribe to the simple "Fido"/Fido alignments of traditional semantic expectation. Putnam labeled these hypothesized correlates as "natural kinds or properties," selecting as his chief paradigm the alleged fact that "possesses a fixed degree of averaged kinetic energy" serves as the "natural kind" to which the term "temperature" eventually gravitates. This enormously influential thesis has become popularly known as "reference magnetism" and is presumed to follow from the alleged truism that "Of course, scientists will want their terminologies to eventually possess unambiguous references." Metaphysically inspired authors have revived antiquated conceptions of scientific methodology by suggesting that Putnam's focusing processes indicate that one of science's primary goals is that of seeking the hidden *essences* of natural states of affairs. Such mystical claims would have sent chills down the spines of our hard-headed nineteenth century avatars.

On this score, let us simply observe that this characterization of "temperature's" evolving career represents a philosopher's myth pure and simple. If it were correct, we couldn't attribute a "temperature" to an ordinary rubber band. But I won't enlarge upon these details here.[18] In any case, Putnam's attempt to codify the

---

[18] See Wilson (2017, 190–2). The motivational complications are intimately connected with the energetic cascades discussed in Chapter 6.

expectations of "scientific realism" in *linguistic terms* should have been rejected as inappropriate from the get-go (there's none of that in Max Planck, for example).

Be this as it may, Putnam's futuristic construal of "scientific realism" in "natural kind" terms opened the floodgates to a revived conception of "metaphysics" that imposes prohibitory restrictions upon scientific innovation closely akin to the tenets that our scientific avatars feared.[19] David Lewis's writings (Lewis 1983) represents a critical stage in these developments, and Sider's remarks follow him closely. Observe how the latter appeals to "scientific realism" in the programmatic declaration that I cited at the end of the previous chapter:

> A realistic picture of science leaves room for a metaphysics tempered by humility.
>
> (Sider 2007, 6)

It is Putnam's linguistic "realism" that is here invoked, not one grounded in the direct consideration of real-life developmental process. For Sider, "realism" demands the eventual production of a "fundamental science" firmly planted upon univocal "natural property" foundations. As already indicated, he shares the conviction that, although we do not possess such an admirable formalism at present, one day we surely will. Because he further presumes that this hypothetical science will conform to the convenient-for-philosophy contours of single-leveled traditional theory T expectation, Sider believes he can safely ignore the "small metaphysics" complexities of ascertaining the objective supports that sustain the utilities of our current employments of scientific vocabularies. Through these swift "idealizations," Sider leapfrogs past all of the inconvenient complexities we have noted in these pages, allowing him to luxuriate in a rosier future free of such concerns. Along the way our philosopher/scientists' original misgivings with respect to "essences," "natural kinds," and allied forms of strong "metaphysical" demand appear to have become utterly forgotten. My suggestion, of course, is that we should attend more warmly to their wise councils.

Although Sider portrays his future-based doctrines as ones of "modest expect-ation," they are not modest at all in the sense that he feels entirely certain of the based categories that must be completed within this hypothesized "fundamental science" (the precise future occupants of these firm categories strike Sider as matters of comparatively little metaphysical interest). That assumption granted, philosophers can busily occupy themselves in constructing a pre-scientific

---

[19] I find it striking that, despite their promises of remaining up to date with science, a considerable segment of the "analytic metaphysics" literature actually struggles with venerable problems within classical mechanics. And they commonly do so with mathematical tools long recognized to be inadequate to these purposes (Smith 2007; Wilson 2008). I once gave a lecture in which I mischievously posted an aerial photograph of a prominent university, pointing out the locations of the departments of philosophy and fluid mechanics. I then asked, "Don't these folks talk to one another?"

"theory" of metaphysical requirement by adhering to the same methodological standards that allegedly characterize rigorous "theory" construction within any designated field. In this imitative fashion, Sider reinstates an essentially apriorist conception of the metaphysical enterprise as a discipline that delineates the contours that any respectable science must later strive to instantiate, largely based upon one's armchair intuitions as to how the required categories of "object," "property," "law," and "cause" might be filled in.

But no reasonable construal of "scientific realism" can promise Sider an all-encompassing "fundamental physics" of the sort he anticipates. Quite the contrary; an objective assessment of the limits that constrain the reach of our available mathematical tools may identify descriptive lacunae that can only be bridged through means other than a single-leveled mathematized formalism. A gambler may huff and puff all he likes, but he won't be able to successfully augur the outcomes of a severely unstable gaming device. And the processes of energetic decoherence in nature may demand similar descriptive liberties.[20]

But in the breezy manner of typical "theory T" appeal, the subtle symptomatology that I have defended as essential to serious conceptual rectification has been cast aside as unimportant. Once these dismissive proclivities take hold, they are extremely hard to dislodge. In my personal experience, I have been dumbfounded by the almost complete indifference to real-life example among Sider's analytic metaphysics cohort: "Oh those doctrinal peculiarities may seem troublesome now, but I needn't worry about any of them. Your worries are merely epistemological." Any suggestion that pursuing "metaphysics" might require a diagnostic probing of linguistic architecture strikes them as a betrayal of the designated mission.

Such are the downstream consequences of the wrong turning that analytic endeavor took in the 1930s, prompted by Carnap's eager embrace of Hertz's hope that confusions of classical mechanics might be remedied through a stiff dose of axiomatic medicine. The fact that the subject did not respond entirely favorably to the treatment occasioned no subsequent alarms within the philosophical community, largely because in the meantime the quantum physicists had lost interest and shunted the patient off to the wards of the departments of engineering. Hertz's and even Carnap's specific proposals have been long forgotten within our contemporary context, and most modern metaphysicians would be highly offended to be accused of seeking axiomatizations in Carnap's fuddy-duddy manner. But these same writers have evidently forgotten virtually all of

---

[20] I might further mention that there is a certain sector of contemporary philosophers of quantum mechanics who refuse to consider any aspect of their subject unless it has been properly axiomatized (for a critique, see Williams (2019)). Like Casey at the bat, they disdain all irregular features such as renormalization and far field asymptotics as "that ain't my style" as the orbs whiz across their plates. They might recall that the mighty Casey struck out through stern adherence to these scruples.

the significant methodological concerns with which our nineteenth century pre-decessors forebears actively struggled, in their attempts to liberate their improving disciplines from the ill-considered manacles of armchair philosophizing. But these same misgivings remain as relevant today as they were back then. I can't see how contemporary analytic effort could have possibly returned to its present-day forms of excessive "metaphysical" presumption without the intervening amnesia of theory T fogginess. And the surface allure of "ersatz rigor" can't obliviate the fact that important considerations have become forgotten along the way. By all means, we should continue to reflect upon our relationships with the exterior world in a "metaphysical" vein, but we should also keep a closer eye upon the range of smaller concerns that generate the grander extravaganzas.

Thus the "alternative history" of analytic philosophy I would have liked to have witnessed would have returned to Mach's and Hertz's struggles with conceptual disharmonies and built upon their recommendations in a more nuanced manner, based upon a careful study of the successful stratagems that we encounter within real-life deductive practice. We have remembered only the dubious remedy, not the informed diagnosis.

9

Truth in a Multiscalar Landscape

[I]t is easy to sit at home and conceive rocks and heath, and waterfalls; and that these journeys are useless labors, which neither impregnate the imagination, nor enlarge the understanding. It is true that of far the greater part of things, we must content ourselves with such knowledge as description may exhibit, or analogy supply; but it is true likewise, that these ideas are always incomplete, and that at least, till we have compared them with realities, we do not know them to be just. As we see more, we become possessed of more certainties, and consequently gain more principles of reasoning, and found a wider basis of analogy. Regions mountainous and wild, thinly inhabited, and little cultivated, make a great part of the earth, and he that has never seen them, must live unacquainted with much of the face of nature, and with one of the great scenes of human existence.

Samuel Johnson (Johnson 1775, 32)

(i)

As previously noted, as soon as a descriptive practice assumes the coordinated contours of multiscalar architecture, a need for *management vocabularies* emerges as well—we must somehow develop terminologies that allow us to specify how a discourse focused upon the ΔL scale length should send suitable corrective messages to one posed at the ΔH scale (and vice versa). This communication task needs to be executed carefully because the two scales often share vocabularies that have become specialized to different forms of referential attachment. Any crude amalgamation of conclusion will result in immediate syntactic contradiction. And a large amount of variable reduction must be accomplished as well to avoid the computational explosions that Tao calls "the curse of dimensionality." Mathematicians call these coordination messages "homogenizations" and probe the intuitive limits involved with extraordinarily delicate precision (E 2011). In ordinary discourse, we must appeal

Imitation of Rigor: An Alternative History of Analytic Philosophy. Mark Wilson, Oxford University Press. © Mark Wilson 2022.
DOI: 10.1093/oso/9780192896469.003.0009

to similar forms of interscalar relationship, albeit in a looser, seat-of-the-pants manner. We do not talk of "homogenizations" in everyday discourse (unless we are applied mathematicians). But we still require some terminological surrogate for addressing the required management tasks. How do we do it?

I've already availed myself frequently of one of these popular surrogates in explicating our architectural arrangements throughout this book, viz., I've employed perceptual metaphors, e.g.:

> Upon the ΔH level, a boundary region slippage at the ΔL will look like a loss of elastic capacity.

However, we are equally likely to employ the terminology of "truth" and "falsity" capture these same requirements on data management:

> If the sentence "The boundary attachments shift in region ρ" becomes true upon the ΔL level, then the sentence "The region ρ loses some of its former elastic capacity" becomes true at the level ΔH.

These simple observations with respect to the language management duties of "truth" strike me as philosophically informative.

The basic genius of a multiscalar routine lies in the fact that its completed recipe eventually unites all of these scale-linked determinations into descriptive harmony via a sequence of successive corrections of the sort we applied to the clothesline of Chapter 6. This tactic requires that the individual feedback corrections are reiterated enough times so that the localized attributions of "true at the scale level ΔL" no longer adjust after repeated corrective feedback from the other scales. Mathematicians say that the sequence of successive approximations has reached a "fixed point" in such a finalized equilibrium.

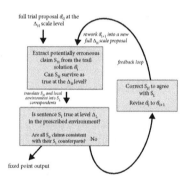

These observations become particularly salient in view of Anil Gupta and Nuel Belnap's "revision theory of truth" which relies upon similar "fixed point" stabilizations of truth value assignment after undergoing a similar progression of evaluative revisions (Gupta and Belnap 1993). The central focus of their work is centered upon the potentially paradoxical patterns of corrective revision that we confront when we attempt to rectify sentences that contain iterated applications of an unparameterized notion of "true" (e.g., in potentially paradoxical claims like "Everything that Saul says is true" where Saul happens to deny various sentences that can potentially undermine the viability of the original claim). However, evaluating a philosophical thesis with respect to the natural language utilities of the term "true" on the basis of such considerations alone is apt to seem unconvincing due to the oddly paradoxical circumstances in which they appear. It strikes me as reassuring to find that closely analogous forms of fixed point stabilization tacitly emerge whenever a multiscalar architecture needs to reach mutually consistent accord amongst its submodels. If my hunch that multiscalar-like architectures are tacitly employed within a wide variety of everyday descriptive practices proves correct, then those same usages will require vocabularies in which the required interscalar messages can be articulated. And this managerial requirement reveals an interesting variety of correspondence truth that has not been adequately explored, to the best of my knowledge.

For reasons I find unconvincing, philosophers frequently reject the notion of "true at the length scale ΔL" as "perspectival" out of hand, as if the evaluation reflects some wuzzy permissiveness beloved of hippies and freshman philosophy students. But as I've already indicated, "perspectival" is a misleading word to employ in these contexts, because the salient evaluative platforms represent descriptive opportunities objectively present in nature, quite independently of whether we recognize their exploitable advantages or not.[1]

However, we must realize that it is also natural to utilize the evaluative term "true" in a wider managerial role operating outside of its current submodel specializations as a means of criticizing the strategic presumptions of interscalar dependency upon which the operative architecture is built. In such a vein, we might observe, "It's simply not true that the main exception to elastic behavior arises when lower scale grain boundaries slip. Temperature effects must be taken into account as well." In doing so, we are claiming that an architecture properly responsible to empirical demand should include an explicit "temperature" parameter as well. We utilize "cause's" underlying managerial capacities in a similar manner when we offer corrections to an inadequately framed explanatory architecture. For example, we might respond to the classic "flashlight paradox" of relativity as follows: "In regarding the sweep of the flashlight as the cause of the faster-than-light projection on the faraway screen, you have mistakenly treated the steady state condition of the light emerging from the flashlight as the 'cause' of the group velocity pattern witnessed on the screen. But this is a mistaken analysis; a proper highlighting of 'cause' should directly attend to the pulses of light as they serially emerge from the flashlight."[2]

---

[1] Richard (2010) and MacFarlane (2014) pursue these themes.

[2] Our everyday talk of light's "causal powers" typically presumes that we are considering a lengthy sequence of close-to-steady-state pulsed events. But such an otherwise convenient policy occasionally generates a seemingly paradoxical clash in evaluating signal "velocity." I have avoided detailed investigation of steady state submodels in this book, but they resemble revised equilibrium treatments in many respects (Wilson 2017, 90–4).

(ii)

The need for managerial vocabularies that can steer language in reformative directions arises as a simple corollary of the evolutionary picture of linguistic development articulated in Chapter 1. By themselves, the silent shaping hands of practical advantage rarely deposit a linguistic usage in a totally perfected state of classificatory or inferential efficiency. Instead, we encounter the same slightly muddied and muddled adaptations that characterize biological developments everywhere. In these respects, the multiscalar architectures we have examined are unusually orderly in their strategic contours, for it is relatively easy to determine the objective qualities with which the word "force" aligns within a selected submodel and to rationalize the reconfigurations of descriptive data that transpire when our computational attention shifts from one length scale to another. As a diagnostic scheme, the proposal offers the philosophical clarity that we seek when we attempt to unravel the formational mysteries of a puzzling practice such as "how do we utilize 'force' with respect to a layered material?" As an object lesson in conceptual examination, I believe that the multiscalar treatment of Hertzian circumstance outlined in this book is immensely illuminating.

But I would not wish to leave readers with the impression that every worthy chain of reasoning can be readily sorted out quite so crisply. The modeling policies that puzzled Hertz had already reached an advanced state of descriptive utility; his problems largely lay in identifying a controlling architecture that was already latently installed within the practices. But the more typical obstacles we encounter with respect to applicational clarity rarely supervene upon such a well-honed strategic basis but instead implement their improvements in more haphazard and less well-marked ways. As I remarked earlier in Wilson (2007), I attempted to map out a reasoning architecture suitable to our everyday use of color words such as "red," but was ultimately defeated by the complexities of the task. By examining the precifications of classification that can be found within professional "color science" (e.g., the various tactics whereby acceptable "colored" products are contractually specified with respect to dyed clothing, automobile sheen, or poly-chromatic film), I was able to discern various platforms of localized usage comparable to the "submodels" utilized within the multiscalar schemes we have discussed. These industrial identifications further suggest that we instinctively employ looser analogs of these specialized criteria within our everyday evaluations of hue (e.g., we examine a dress near a window or inspect the car from different viewing angles). Based upon these applicational hints and with some knowledge of their physical and psychological bases, I could determine that a word like "red" somehow swiftly adjusts its referential attachments in manners that secretly reflect the localized opportunities required for practical achievement. So I felt that I had obtained a preliminary appreciation of the tangled threads of informational transfer and reconfiguration that bind these descriptive patches together into the well-adapted unities of everyday usage, comparable to the "homogenization messages" central to the efficiencies of our multiscalar architectures.

But I remained fully aware that these observations merely comprise the beginning steps in an adequate diagnosis of the real-life complexities of color vocabulary. In the final analysis, I could only supply somewhat hazy intimations of the rich underlying architecture that maintains our color talk in a condition of effective but complex coherence. Doing better would have required a vastly greater command of applicational nuance than I possess, either then or now. This undeveloped state left many of my readers dissatisfied, as if I had hidden my lack of philosophical convictions behind a screen of illustrative vagueness. But I had hidden nothing, for my "philosophical convictions" merely reflect an inclination to approach the puzzles of real-life conceptual confusion from the perspective of evolutionary shaping and "scientific realism." At this highest level of generality, we can scarcely articulate any informative claim beyond "language development does some pretty funny things in response to the practical pressures of objective strategic opportunity." But when we descend to particular instances, we find that the devoted labors of applied mathematicians have already provided the basic tools for unraveling the relatively orderly adjustments that generate the conceptual oddities that confused Hertz. So this is the explanatory setting that I have exploited throughout this book. I cannot offer anything comparatively crisp with respect to the behaviors of our color words, for we remain largely in the dark with respect to many of the shaping considerations that explicate why our color words behave as strangely as they do. Nonetheless, our experiences with classical "force" offer a preliminary appreciation of some of the basic architectural interconnections that will surely be involved.

Upon this basis alone, we can respond to old-fashioned philosophical questions such as "Do color words reflect objective features of the world?" in a helpful and "scientific realist" manner. "Yes," we can reply, "but these reports arise within usages whose architectural scaffoldings must be recognized and understood before the rationales for the localized registrations of objective data can be properly identified." When this goal is finally achieved, we will be able to address Joseph Addison's disquieting metaphor in a more suitable way:

Things would make but a poor appearance to the eye, if we saw them only in their proper figures and motions. And what reason can we assign for their exciting in us many of those ideas which are different from anything that exists in the objects themselves (for such are light and colors), were it not to add supernumerary ornaments to the universe, and make it more agreeable to the imagination? ... In short, our souls are at present delightfully lost and bewildered in a pleasing delusion, and we walk about like the enchanted hero of a romance, who sees beautiful castles, woods, and meadows; but upon the finishing of some secret spell, the fantastic scene breaks up, and the disconsolate knight finds himself on a barren heath, or in a solitary desert.

(Addison 1717, 334)

In truth, if Addison's knight were to awaken with a suitable appreciation of the strategic dimensions of color experience, he would find that his "beautiful castles, woods, and meadows" remain entirely intact but have become subsequently enveloped within a richer netting of descriptive strategy and sensory capacity. I like to compare this enlightened condition to that of Heinrich Hertz himself, observing feeble sparks between induction coils in a cramped laboratory and realizing that they portend the presence of a vastly complicated hidden world of electromagnetics that enfolds every aspect of our daily lives.[3]

In essence, this book's complaints with respect to "ersatz rigor" begin here. The philosophical dilemmas that we commonly confront in struggling with an established usage trace to the imperfect agglomerations characteristic of an improving practice, and these obstacles may stem from an extremely wide set of applicational subtleties. A philosophical sleuth must begin with the misty symptoms of conceptual irregularity as they organically pop up here and there within a developing usage. She should not expect that these tensions will submit to any common pattern of diagnostic resolution but are apt to prove as potentially varied as nature itself. So it is truly a happy occasion when we learn that the mysterious policies we confronted in Physics 101 long ago can be effectively rationalized through the wonderful efficiencies of a multiscalar architecture. But it is comparatively rare that an adequate analysis can be neatly fit within these relatively tidy confines, let alone sandwiched within the even more straitened contours of an axiomatic scheme. To be sure, formal precision in architectural arrangement is often desirable in the concluding stages of some long-persisting conceptual mystery, as when we hope to implement our improved understanding of "force's" applicational wiles upon a computer. Nonetheless, the core of the hard intellectual work needs to come before, in identifying the practical advantages offered within the confusing complexities of naturally evolved practice and figuring out how information gets processed amongst the resulting patches. In this manner, the tasks of a philosophical detective often resemble those of the professional mathematician. The "true mathematics" generally emerges as the latter struggles with doodles on coffee shop napkins, attempting to discern some deeper pattern within her scribbles. Writing up her insights as "proofs" in a professional way afterward constitutes an important verificatory chore, but it rarely captures the locale where her deepest insights were achieved.

---

[3] As we noted in Chapter 8, section (ii), our favored philosopher/scientists generally failed to confront the doctrines of direct mental acquaintance situated at the base of Addison's eerie metaphor.

Our intellectual predicament demands that philosophical inquiry must meander along messy and shambling paths, in the manner of Peter Falk's television detective Columbo. But the preponderance of methodological opinion within current philosophical endeavor has instead drifted towards two opposed poles of ill-conceived precisification, both of which take their roots within our late nineteenth-century era (an excellent survey of these attitudes is Daly (2010)). The first of these grounds its projects within allegedly well-established "meanings," as revealed to us in the "intuitions" we frame with respect to anticipated applications. A previous work (Wilson 2006) supplies lengthy critique of the "classic picture of meaning" upon which such views rely, which I have not repeated here. But the reverse side of the methodological coin traces to the heritage reviewed in this work: intuitive "meanings" and "references" comprise the metaphysical fictions of the projective mind whose distortions a philosopher should rectify through quasi-axiomatic "rational reconstructions" of a Carnapian character (Carus 2007). This point of view elevates a inflexible conception of absolutist "rigor" to a prominence that I have rejected as inimical to the philosophical enterprise. Implicit in these wayward developments is the presumption that a philosopher only needs to attend to the coarse "logical structure of a (mythically completed) theory," without investigating how the inferential pathways of practical utility actually run in real life.[4] And we've seen that these distortions of "ersatz rigor" linger on, even as recent philosophers return to the verdant fields of speculative metaphysics. But to conceptualize the philosophical enterprise in any of these ways renders an injustice to the wide-ranging suppleness of thought required in its real-life detective work. The counterfactual "alternative history" to which our subtitle refers represents the manner in which analytic philosophy might have developed, had it remained more faithful to the original insights of Hertz and Mach.

(iii)

To successfully refurbish our ships while they are yet sailing upon the high seas (Neurath 1921) requires *managerial vocabularies* of an appropriate capacity, as we have already observed with respect to "cause" and "true." But the ordinary language term "refers" commonly plays a comparable orientational role with respect to proper nouns and predicates. Thus we might warn a novice, "When we drop to a lower length scale in describing a viscous fluid, recognize that a

---

[4] I think that it is reasonably likely that the latter Wittgenstein extracted methodological morals from Hertz in some manner akin to that I have followed here, but his writings are drenched in the bathetic colors of "therapy" and "philosophical suffering." I want none of those trappings here; as Wilson (2006, 614) proudly proclaims, "I yield the lamp of scientism to no one."

portion of the 'force' designated upon a macroscopic scale shifts instead to becoming part of the referent of 'molecular movement'." As we've seen, such adjustments in localized reference can be deftly handled by sending suitably "homogenized" messages between the length scales, which prevent these differences in local references from creating significant problems. In this sense, the underlying "logic" of a successful reasoning scheme may secretly embody a long sequence of rather subtle informational adjustments.

Accordingly, terms like "refers" and "meaning" should be regarded as the everyday vernacular in which we diagnose and recalibrate our improving usages (but they service other diagnostic duties as well, as Courant indicates in his discussion of a mathematician's specialized interest in "existence"). We must first map out the roughly hewn informational correlations upon which the usage is founded and the strategies of contextualized transference that underpin its inferential successes (this diagnosis is often the hardest part, as already suggested). We can then attempt to repair any lingering lapses with systematic repairs. As such, consideration of a word's "reference" typically combines descriptive reporting with a fair dollop of reformatory suggestion. As we have noted, analytic philosophers enamored of "intuitions" have elevated these quasi-reformatory and often roughly hewn assessments ("temperature refers to mean kinetic energy") into mythical totems of a suspiciously rigidified character. But we should not follow their lead. For essentially the same reasons, we should also reject the self-styled "deflationists" who regard "meaning" and "reference" talk as unconcerned with real linkages between words and world and merely implement weak syntactic advantages such as those attaching to what Quine calls "semantic assent" (Quine 1953). Our appreciation of linguistic improvement should be expected to aperiodically wobble between diverse "seasonalities" of focused attention, sometimes concentrating upon architectural arrangement and sometimes focusing upon informational registration.

It's not *language's* fault that philosophers frequently picture the word-to-world attachments of its successful terminologies in distorted or overly simplified manners ("our pencils are often wiser than ourselves"). Our proper philosophical duty is to unmask the successful informational alignments as best we can, while simultaneously discarding the simplified pictures that we may have previously entertained with respect to their informational ministrations. With sufficient investigative diligence, we can eventually improve upon the wisdoms our pencils have bequeathed us, usually through gentle tweaking rather than roughly rejecting their councils in ill-considered axiomatic refashionings (which often represent the linguistic analogs of what urban developers did to Pennsylvania Station). Wiser upgradings will strive to retain most of the grand old edifice, but with a considerably revised appreciation of its structural interdependencies.

Conceptual quandaries can arise within every walk of life, and anyone who can help straighten them out should qualify as a "philosopher" (lots of impressive mathematical work strikes me as of this character). By cultivating a broad awareness of how allied dilemmas have emerged within an array of subjects, a free-thinking diagnostician can contribute significantly to improving our intellectual circumstances by helping to unravel these unavoidable tangles. But this task often requires a command of a wide-ranging detail that may appear irrelevant to the original puzzles at first glance. In consequence, one should never dismiss any potential stream of data out of hand as "irrelevant" on a priori grounds.

Thus I believe that philosophers should fancy themselves as dedicated students of the irregular. But inherited fashions within our subject presently reward ersatz rigor and armchair pontification without encouraging the informed attention to detail and practical application upon which proper conceptual disentanglements characteristically depend. By examining a single historical example as an illustrative test case in this book, I have attempted to demonstrate why apriorist intellectual shackles of these blinkered kinds are unlikely to produce a satisfactory resolution of the original quandaries.

I believe that such morals extend to wider areas of philosophy than the physics-centered examples addressed here. Thus, an adequate understanding of the acts we call "lies" must surely attend to the architectural surroundings of such deeds. Did shopkeeper Jones behave dishonorably to customer Smith when Jones said X to Smith? On first blush, X might appear to represent a straightforward factual fib, but, in light of the corrective safeguards that prevail within the encompassing society, perhaps not. I here allude to J. L. Austin's celebrated essay of "excuses" (Austin 1966), for his methodologies with respect to contextual detail do not seem altogether disjoint from our own.[5]

All in all, I believe that we will be better served if we imitate the intellectual modesties characteristic of the British philosophers, poets, and essayists of the seventeenth and eighteen centuries who advise us to commence with a:

> Survey of our own Understandings, examine our own Powers, and see to what Things they were adapted. Till that was done, I suspected that we began at the wrong end, and ... loose our Thoughts into the vast Ocean of Being, as if all the boundless Extent, were the natural and undoubted Possessions of our Understandings, wherein there was nothing that escaped its Decisions, or that escaped its Comprehension. Thus Men, extending their Enquiries beyond their Capacities, and letting their Thoughts wander into those depths where they can find no sure Footing.   (Locke 1689, 47)

---

[5] As I remark in Wilson (2017, 5), Austin's diagnostic skills appear to be significantly out of fashion nowadays. It is worth noting that a contemporary comrade in "scientistic philosophy" (Maddy 2017) expresses allied affinities with his endeavors.

Or we can likewise cite Joseph Glanvill in warning us to not rest our self-estimations upon the unattainable vanities of dogmatic certainty:

> Neither are we yet at so deplorable a loss, in the other parts of what we call Science; but that we may meet with what will content ingenuity, at this distance from perfection, though all things will not completely satisfy strict and rigid inquiry. Philosophy indeed cannot immortalize us, or free us from the inseparable attendants on this state, Ignorance and Error. But shall we malign it, because it entitles us not to an Omniscience?   (Glanvill 1665, 206)[6]

The cheerier codicil I would append to these bleak assessments of human attainment is one that highlights the novel forms of conceptual bootstrapping that allow us to adapt old words to novel purposes, through the "seasonal" adjustments of linguistic policy that we have here surveyed. Perhaps the resulting point of view might be dubbed as a *mitigated realism*,[7] in acknowledgment of its emphases upon the manners in which we cobble together unexpectedly roundabout strategies in adapting our native reasoning tools (including mathematics) to the recalcitrant universe in which we find ourselves.

In this spirit, let me conclude with a final appeal to that redoubtable avatar of unexpected opportunity (Oliver Heaviside) in his struggles against the "ersatz rigorists" of his day:

> This is fudge and fiddlesticks. . . . I know mathematical processes that I have used with success for a very long time, of which neither I nor anyone else understands the scholastic logic. I have grown into them and so understand them that way. Facts are facts, even though you cannot see your way to a complete theory of them. And no complete theory is possible. There is always something wanting, no matter how logical people may pretend otherwise.
>
> (Heaviside 1912, III 515)

Heaviside

---

[6] The impermanence of firm knowledge is aptly illustrated by the fact that Glanvill was a stalwart defender of witchcraft and directly influenced America's Cotton Mather.

[7] Williams (2021) labels this perspective as "non-rigid realism."

Bibliography

Addison, Joseph 1717, *The Works of Joseph Addison*, Vol. VI (New York: Putnam, 1854).

Armstrong, David 1983, *What Is a Law of Nature?* (Cambridge: Cambridge University Press).

Austin, J. L. 1966, "A Plea for Excuses" in *Philosophical Papers* (Oxford: Oxford University Press).

Bacon, Andrew 2020, "Logical Combinatorialism," *Philosophical Review* 129(4).

Baird, David, Hughes, R. I. G., and Nordmann, Alfred, eds. 1998, *Heinrich Hertz: Classical Physicist, Modern Philosopher* (Dordrecht: Kluwer).

Batterman, Robert 2013, "The Tyranny of Scales" in Robert Batterman, ed., *The Oxford Handbook of Philosophy of Physics* (Oxford: Oxford University Press).

Batterman, Robert 2021, *In Middle Way* (Oxford: Oxford University Press).

Baz, Avner 2012, *When Words Are Called For* (Cambridge, MA: Harvard University Press).

Bloch, A.M. 2003, *Nonholonomic Mechanics and Control* (New York: Springer).

Boltzmann, Ludwig 1897, "On the Indispensability of Atomism in Natural Science" in *Theoretical Physics and Philosophical Problems* (Dordrecht: D. Reidel, 1974).

Bonawitz, E. B., Ferranti, D., Saxe, R., Gopnik, A., Meltzoff, A. N., Woodward, J., and Schulz, L. E. 2010, "Just Do It? Investigating the Gap between Prediction and Action in Toddlers' Causal Inferences," *Cognition* 115(1).

Boussinesq, Joseph 1880, *Leçons synthétiques de Mécanique générale*. Quoted in Duhem (1914).

Boyle, Robert 1674, "Of the Excellency and Grounds of the Corpuscular or Mechanical Philosophy" in *Selected Philosophical Papers of Robert Boyle* (Indianapolis: Hackett, 1991).

Brogliato, Bernard 2016, *Nonsmooth Mechanics* (Switzerland: Springer).

Brooker, Geoffrey 2003, *Modern Classical Optics* (Oxford: Oxford University Press).

Carnap, Rudolf 1923, "On the Task of Physics" in *Rudolf Carnap: Early Writings* (Oxford: Oxford University Press, 2019).

Carnap, Rudolf 1932, "Elimination of Metaphysics through Logical Analysis of Language," Arthur Pap, trans. in A. J. Ayer, ed., *Logical Positivism* (Glencoe, IL: The Free Press, 1959).

Carnap, Rudolf 1934, *The Logical Syntax of Language* (London: Routledge and Kegan Paul, 1959).

Carnap, Rudolf 1947, *Meaning and Necessity* (Chicago: University of Chicago Press).

Carnap, Rudolf 1948, *Introduction to Semantics* (Cambridge, MA: Harvard University Press).

Carus, A. W. 2007, *Carnap and Twentieth-Century Thought* (Cambridge: Cambridge University Press).

Cassirer, Ernst 1950, *The Problem of Knowledge* (New Haven: Yale University Press).

Chalmers David, Manley, David, and Wasserman, Ryan 2009, *Metametaphysics* (Oxford: Oxford University Press).

Clifford, William Kingdon 1901, *Lectures and Essays*, Vol. II (London: MacMillan and Co.).

Coleridge, S. T. 1817, *Biographia Literaria* (Princeton: Princeton University Press 1983).

Coriolis, Gaspard-Gustave 2005, *Mathematical Theory of Spin, Friction and Collision in the Game of Billiards*, David Nadler, trans. (San Francisco: Nadler Publications).

Corry, Leo 2010, *David Hilbert and the Axiomatization of Physics* (Berlin: Springer).

Courant, Richard 1950, *Dirichlet's Principle, Conformal Mapping and Minimal Surfaces* (New York: Interscience).

Daly, Chris 2010, *An Introduction to Philosophical Methods* (Toronto: Broadview).

de Boer, Reint 2000, *Theory of Porous Media: Highlights in Historical Development and Current State* (Berlin: Springer).

Dedekind, Richard 1854, "On the Introduction of New Functions in Mathematics" in *From Kant to Hilbert, vol. 2*, William Ewald, ed. (Oxford: Oxford University Press, 1996).

Descartes, René 1628, "Rules for the Direction of the Mind" in *The Philosophical Works of Descartes*, Elizabeth S. Haldane and G. R. T. Ross, trans. (Cambridge: Cambridge University Press, 1968).

Descartes, René 1641, "Meditations on First Philosophy" in *The Philosophical Writings of Descartes*, Vol. II (Cambridge: Cambridge University Press, 1985).

Destrade, Michel, Murphy, Jeremiah, and Saccomandi, Giuseppe 2019, "Rivlin's Legacy in Continuum Mechanics and Applied Mathematics," *Philosophical Transactions of the Royal Society* A 377.

Dingle, Robert B. 1973, *Asymptotic Expansions: Their Derivation and Interpretation* (London: Academic Press).

Dowe, Phil 2000, *Physical Causation* (Oxford: Oxford University Press).

Du Châtelet, Emilie 1742, "Foundations of Physics" in *Selected Philosophical and Scientific Writings*, J. P. Zinsser and I. Bour, trans. (Chicago: University of Chicago Press, 2009).

Duhem, Pierre 1903, *The Evolution of Mechanics*, J. M. Cole, trans. (Alphen aan den Rijn: Springer, 1980).

Duhem, Pierre 1914, *The Aim and Structure of Physical Theory*, Philip Weiner, trans. (Princeton: Princeton University Press, 1954).

E, Weinan 2011, *Principles of Multiscale Modeling* (Princeton: Princeton University Press).

Eddington, A. S. 1928, *The Nature of the Physical World* (London: MacMillan and Co.).

Enriques, Federico 1914, *The Problems of Science* (Chicago: Open Court).

Euler, Leonhard 1768, *Letters of Euler on Different Subjects in Natural Philosophy Addressed to a German Princess I*, David Brewster, trans. (New York: Harper and Brothers, 1854).

Field, Hartry 2001, *Truth and the Absence of Fact* (Oxford: Oxford University Press).

Field, Hartry 2003, "Causation in a Physical World" in Michael J. Loux and Dean W. Zimmerman (eds.), *The Oxford Handbook of Metaphysics* (Oxford: Oxford University Press).

Frege, Gottlob 1914, "Logic in Mathematics," in *Posthumous Writings* (Oxford: Blackwell's, 1991).

Gardner, Martin 1952, *In the Name of Science* (New York: Putnam). Republished by Dover in 1957 as *Fads and Fallacies in the Name of Science*.

Glanvill, Joseph 1665, *Scepsis Scientifica; or, the Vanity of Dogmatizing* (London: Kegan Paul and Trench, 1885 reprint).

Goldsmith, Werner 2001, *Impact* (New York: Dover).

Gupta, Anil 2011, *Truth, Meaning, Experience* (New York: Oxford University Press).

Gupta, Anil and Nuel Belnap 1993, *The Revision Theory of Truth* (Cambridge: MIT Press).

Gurtin, Morton 1999, *Configurational Forces as Basic Concepts of Continuum Physics* (New York: Springer).

Hadamard, Jacques 1903, *Leçons sur la propagation des ondes et les équations de l'hydro-dynamique* (Paris: Hermann).

Hadamard, Jacques 1923, *Lectures on Cauchy's Problem in Linear Partial Differential Equations* (New Haven: Yale University Press).

Hamel, Georg 1921, *Grundbegriffe der Mechanik* (Leipzig: B. G. Teubner).

Heaviside, Oliver 1912, *Electromagnetic Theory, I–III* (Providence: Chelsea, 2003).

Helm, Georg 1898, *The Historical Development of Energetics*, R. J. Deltete, trans. (Dordrecht: Kluwer, 2000).

Helmholtz, Hermann von 1878, "The Facts in Perception" in *Epistemological Writings*, Malcolm F. Lowe, trans. (Dordrecht: D. Reidel, 1977).

Hempel, Carl 1970, *Aspects of Scientific Explanation* (New York: Free Press).

Hertz, Heinrich 1881, "On the Contact of Elastic Solids" in *Miscellaneous Papers*, D. E. Jones and G. A. Schott, trans. (London: MacMillan and Co., 1896).

Hertz, Heinrich 1889, "Light and Electricity" in *Miscellaneous Papers*, D. E. Jones and G. A. Schott, trans. (London: MacMillan and Co., 1896).

Hertz, Heinrich 1891, "Hermann von Helmholtz" in *Miscellaneous Papers*, D. E. Jones and G. A. Schott, trans. (London: MacMillan and Co., 1896).

Hertz, Heinrich 1893, *Electric Waves*, D. E. Jones, trans. (New York: Dover, 1962).

Hertz, Heinrich 1894, *The Principles of Mechanics*, D. E. Jones and J. T. Walley, trans. (New York: Dover, 1956).

Hilbert, David 1902, "Mathematical Problems," Mary Winston Newson, trans., *Bulletin of the American Mathematical Society* 8.

Hilbert, David 1905, "Lecture Notes," cited in Michael Janssen, "Arches and Scaffolds" in Love and Wimsatt, eds., *Beyond the Meme* (Minneapolis: University of Minnesota Press, 2019).

Hilbert, David 1925, "On the Infinite," *Mathematische Annalen* 95(2).

Hume, David 1740, "An Abstract of . . . a Treatise" in David Fate Norton and Mary J. Norton, eds., *A Treatise of Human Nature*, Vol. 1: *Texts* (Oxford: Clarendon Press, 2011).

Hume, David 1748, *An Enquiry Concerning Human Understanding* (Oxford: Oxford University Press, 1998).

James, William 1884, "The Dilemma of Determinism," *Unitarian Review and Religious Magazine* XXII.

Johnson, Samuel 1775, *A Journey to the Western Islands of Scotland* (Oxford: Oxford University Press, 2021).

Kant, Immanuel 1786, *Metaphysical Foundations of Natural Science*, Michael Friedman, trans. (Cambridge: Cambridge University Press, 2004).

Kant, Immanuel 1787, *Critique of Pure Reason*, Norman Kemp Smith, trans. (New York: St. Martin's, 1929).

Keller, Joseph B. 1959, "Large Amplitude Motion of a String," *American Journal of Physics* 27(584).

Kennedy, Graeme J. and Martins, Joaquim 2012, "A Homogenization-Based Theory for Anisotropic Beams with Accurate Through-Section Stress and Strain Prediction," *International Journal of Solids and Structures* 49(1).

Kuhn, Thomas 1962, *The Structure of Scientific Revolutions* (Chicago: University of Chicago Press).

Lamb, Horace 1923, *Dynamics* (Cambridge: Cambridge University Press).

Laplace, Pierre Simon 1814, *A Philosophical Essay on Probabilities*, F. W. Truscott and F. L. Emory, trans. (New York: Dover, 1951).

Leitgeb, Hannes and Carus, André 2020, "Tolerance, Metaphysics, and Meta-Ontology," *The Stanford Encyclopedia of Philosophy*: https://plato.stanford.edu/entries/carnap/tolerance-metaphysics.html.

Lewis, David 1983, "New Work for a Theory of Universals," *Australasian Journal of Philosophy* 61(4).

Locke, John 1689, *An Essay Concerning Human Understanding*, P. H. Nidditch, ed. (Oxford: Clarendon Press, 1975).

Lützen, Jesper 2005, *Mechanistic Images in Geometric Form* (Oxford: Oxford University Press).

McKinsey, J. C. C., Sugar, A. C., and Suppes, Patrick 1953, "Axiomatic Foundations of Classical Particle Mechanics," *Journal of Rational Mechanics and Analysis* 2.

MacFarlane, John 2014, *Assessment Sensitivity* (Oxford: Oxford University Press).

Mach, Ernst 1882, "The Economical Nature of Physical Theory" in *Popular Scientific Lectures*, T. J. McCormack, trans. (LaSalle: Open Court, 1895).

Mach, Ernst 1883, *The Science of Mechanics*, T. J. McCormack, trans. (Chicago: Open Court, 1919).

Mach, Ernst 1896, *Principles of the Theory of Heat*, T. J. McCormack and P. E. B. Jourdain, trans. (Dordrecht: D. Reidel, 1986).

Mach, Ernst 1909, *History and Root of the Principle of the Conservation of Energy*, Philip E. B. Jourdain, trans. (Chicago: Open Court, 1911).

Maddy, Penelope 2007, *Second Philosophy* (Oxford: Oxford University Press).

Maddy, Penelope 2017, *What Do Philosophers Do?* (Oxford: Oxford University Press).

Maddy, Penelope 2022, *A Plea for Natural Philosophy and Other Essays* (New York: Oxford University Press).

Mahon, Basil 2017, *The Forgotten Genius of Oliver Heaviside* (Amherst, NY: Prometheus).

Marlow, Wayland C. 1995, *The Physics of Pocket Billiards* (Palm Beach Gardens: Marlow Advanced System Technologies).

Maudlin, Tim 2007, *The Metaphysics within Physics* (New York: Oxford University Press).

Maugin, Gérard A. 1998, *Thermodynamics of Nonlinear Irreversible Behaviors* (Singapore: World Scientific).

Maugin, Gérard A. 2011, *Configurational Forces* (Boca Raton: CRC Press).

Maugin, Gérard A. 2017, *Non-Classical Continuum Mechanics* (Singapore: Springer).

Maxwell, J. C. 1878, "Atom" in *The Scientific Papers of James Clerk Maxwell* (New York: Dover, 1965).

Maxwell, J. C. ND, "In Memory of Edward Wilson": https://mypoeticside.com/poets/james-clerk-maxwell-poems.

Mayr, Ernst 1997, *Evolution and the Diversity of Life* (Cambridge, MA: Harvard University Press).

Monna, A. F. 1975, *Dirichlet's Principle* (Utrecht: Oosthoek, Scheltema, and Holkema).

Murakami, Sumio 2012, *Continuum Damage Mechanics* (Dordrecht: Springer).

Nagel, Ernst 1961, *The Structure of Science* (New York: Harcourt, Brace, and World).

Neurath, Otto 1921, "Anti-Spengler" in *Empiricism and Sociology* (Dordrecht: D. Reidel, 1973).

Nordmann, Alfred 1998, "Everything Could Be Different" in Baird, Hughes, and Nordmann (1998).

Olson, Richard 1975, *Scottish Philosophy and British Physics 1750–1880* (Princeton: Princeton University Press).

Papastavridis, J. G. 2002, *Analytical Mechanics* (Oxford: Oxford University Press).

Paul, L. A. 2012, "Metaphysics as Modeling: The Handmaiden's Tale," *Philosophical Studies* 160(1).

Paul, L. A. and Hall, Ned 2013, *Causation: A User's Guide* (Oxford: Oxford University Press).

Pearl, Judea 2009, *Causation: Models, Reasoning and Inference* (Cambridge: Cambridge University Press).

Pearson, Karl 1892, *The Grammar of Science* (London: Walter Scott).

Peirce, C. S. 1976, *The New Elements of Mathematics* I (The Hague: Mouton).

Pincock, Christopher 2010, "Critical Notice: Mark Wilson, Wandering Significance," *Philosophia Mathematica* 18.

Pippin, Robert B. 2012, *Hollywood Westerns and American Myth* (New Haven: Yale University Press).

Pippin, Robert B. 2021, *Douglas Sirk, Filmmaker and Philosopher* (New York: Bloomsbury).

Planck, Max 1932, *The Mechanics of Deformable Bodies*, Henry L. Brose, trans. (London: MacMillan and Co.).

Podio-Guidugli, Paolo 2019, *Continuum Thermodynamics* (Switzerland: Springer).

Poincaré, Henri 1905, *Science and Hypothesis*, George Bruce Halsted, trans. (New York: The Science Press).

Pope, Alexander 1948, *Selected Works* (New York: Random House).

Poston, Tim and Stewart, Ian 1978, *Catastrophe Theory* (London: Pitman).

Potter, Michael 2020, *The Rise of Analytic Philosophy* (London: Routledge).

Poynting, J. H. 1899, "Address to the Mathematical and Physical Section of the British Association for the Advancement of Science," *Science*, New Series, 10(247).

Putnam, Hilary 1975, "The Meaning of 'Meaning'" in *Mind, Language and Reality* (Cambridge: Cambridge University Press, 1979).

Putnam, Hilary 1978, *Meaning and the Moral Sciences* (London: Routledge and Kegan Paul).

Quine, W. V. 1953, *From a Logical Point of View* (New York: Harper and Row).

Quine, W. V. 1969, *Ontological Relativity and Other Essays* (New York: Columbia University Press).

Rayleigh, Lord 1900, "On Approximately Simple Waves," *Philosophical Magazine*, Series 5, 50(302).

Reichenbach, Hans 1958, *The Philosophy of Space and Time* (New York: Dover).

Reid, Constance 1991, "Hans Lewy 1904–1988" in P. Hilton, F. Hirzebruch, and R. Remmert, eds., *Miscellanea Mathematica* (Berlin: Springer).

Richard, Mark 2010, *When Truth Gives Out* (Oxford: Oxford University Press).

Rivlin, R. S. 1997, "Red Herrings and Sundry Unidentified Fish in Nonlinear Continuum Mechanics" in *Collected Papers*, Vol. 2 (New York: Springer).

Rozikov, Utkir A. 2019, *An Introduction to Mathematical Billiards* (Singapore: World Scientific).

Russell, Bertrand 1914, *Our Knowledge of the External World* (Chicago and London: Open Court).

Russell, Bertrand 1918, "On the Notion of Cause" in *Mysticism and Logic* (London: George Allen and Unwin).

Russell, Bertrand 1940, *An Inquiry into Meaning and Truth* (London: George Allen and Unwin).

Ryle, Gilbert 1953, "Ordinary Language," *The Philosophical Review* 62(2).

Salmon, Wesley 1984, *Scientific Explanation and the Causal Structure of the World* (Princeton: Princeton University Press).

Scheffler, Israel 1963, *The Anatomy of Inquiry* (New York: Knopf).

Schlosshauer, Maximilian 2010, *Decoherence and the Quantum-to-Classical Transition* (Berlin: Springer).

Scott, Wilson 1970, *The Conflict between Atomism and Conservation Theory* (London: MacDonald).

Shaw, F. S. 1953, *An Introduction to Relaxation Methods* (New York: Dover).

Shearer, Michael and Levy, Rachel 2015, *Partial Differential Equations* (Princeton: Princeton University Press).

Sider, Theodore 2007, "Introduction" in T. Sider, J. Hawthorne, and D. Zimmerman, eds., *Contemporary Debates in Metaphysics* (Hoboken: Wiley-Blackwell).

Sider, Theodore 2014, *Writing the Book of the World* (Oxford: Oxford University Press).

Simon, Herbert 1954, "The Axiomatization of Classical Mechanics," *Philosophy of Science* 21(4).

Sklar, Lawrence 2013, *Philosophy and the Foundations of Dynamics* (Cambridge: Cambridge University Press).

Smith, Henry J. S. 1851, "'On Some of the Methods at Present in Use in Pure Geometry'" in *Collected Mathematical Papers* I (Bronx: Chelsea, 1979).

Smith, Sheldon 2000, "Resolving Russell's Anti-Realism about Causation," *The Monist* 83(2).

Smith, Sheldon 2007, "Continuous Bodies, Impenetrability, and Contact Interactions," *The British Journal for the Philosophy of Science* 58(3).

Sommerfeld, Arnold 1943, *Mechanics*, Martin O. Sheen, trans. (London: Academic, 1964).

Spivak, Michael 2010, *Physics for Mathematicians: Mechanics* 1 (Houston: Publish-or-Perish, 2010).

Stebbing, L. Susan 1937, *Philosophy and the Physicists* (London: Methuen).

Steiner, Jakob 1832, *Systematische Entwicklung der Abhängigkeit Geometrischer Gestalten von Einander*, cited in R. E. Moritz, *Memorabilia Mathematica* (New York: MacMillan and Co., 1914).

Stronge, W. J. 2004, *Impact Mechanics* (Cambridge: Cambridge University Press).

Sylvester, James Joseph 1875, "History of the Plagiograph," *Nature* 12(168).

Tait, P. G. 1890, *Properties of Matter* (Edinburgh: Adam and Charles Black).

Tait, P. G. 1895, *Heat* (London: MacMillan and Co.).

Tao, Terence 2012, "E pluribus unum: From Complexity, Universality," *Daedalus* 141(3).

Thomson, William (Lord Kelvin) 1901, "Nineteenth-Century Clouds over the Dynamical Theory of Heat and Light," *Philosophical Magazine* 2(7).

Thomson, William (Lord Kelvin) 1904, Baltimore Lectures on Molecular Dynamics and the Wave Theory of Light (London: C. J. Clay and Sons).

Thomson, William (Lord Kelvin) and Tait, P. G. 1912, *Treatise on Natural Philosophy*, two vols. (Cambridge: Cambridge University Press).

Truesdell, Clifford 1968, "The Creation and Unfolding of the Concept of Stress" in *Essays in the History of Mechanics* (Berlin: Springer).

Truesdell, Clifford 1977, *A First Course in Rational Continuum Mechanics* (London: Academic).

Truesdell, Clifford 1984, "Suppsian Stews" in *An Idiot's Fugitive Essays on Science* (Berlin: Springer).

van Fraassen, Bas 1980, *The Scientific Image* (Oxford: Clarendon Press).

Vardoulakis, Ioannis 2019, *Cosserat Continuum Mechanics* (Cham: Springer).

Voltaire 1901, "Jeannot and Colin," *The Works of Voltaire*, Vol. 3, W. F. Fleming, trans. (Paris: E. R. DuMont, 1901).

Whewell, William 1847, *An Elementary Treatise on Mechanics* (Cambridge: Deightons, Whittaker, and Co.).

Wightman, A. S. 1983, "Hilbert's Sixth Problem" in F. E. Browder, ed., *Mathematical Developments Arising from Hilbert Problems* (Providence: American Mathematical Society).

Williams, Porter 2019, "Scientific Realism Made Effective," *The British Journal for the Philosophy of Science* 70(1).

Williams, Porter 2021, "Realism without Rigidity?," *Philosophy and Phenomenological Research*, forthcoming.

Williamson, Timothy 2008, *The Philosophy of Philosophy* (Oxford: Blackwell's).

Wilson, Mark 2006, *Wandering Significance* (Oxford: Oxford University Press).

Wilson, Mark 2008, "Beware of the Blob: Cautions for Would-Be Metaphysicians" in Dean Zimmerman, ed., *Oxford Studies in Metaphysics 4* (Oxford: Oxford University Press).

Wilson, Mark 2009, "Determinism: The Mystery of the Missing Physics," *British Journal for the Philosophy of Science* 60(1).

Wilson, Mark 2010, "Frege's Mathematical Setting" in Tom Ricketts and Michael Potter, eds., *A Cambridge Companion to Frege* (Cambridge: Cambridge University Press).

Wilson, Mark 2013, "What is 'Classical Mechanics' Anyway?" in Batterman (2013).

Wilson, Mark 2017, *Physics Avoidance* (Oxford: Oxford University Press).

Wilson, Mark 2020, *Innovation and Certainty* (Cambridge: Cambridge University Press).

Wilson, Mark 2021, "Newton in the Pool Hall," in Eric Schliesser and Chris Smeenk, eds., *The Oxford Handbook of Newton* (Oxford: Oxford University Press).

Wilson, Mark 2022, "The Evil Deceiver Strikes Again!," forthcoming in *The Australasian Journal of Philosophy*.

Wolf, Emil 2007, *Introduction to the Theory of Coherence and Polarization of Light* (Cambridge: Cambridge University Press).

Woodhouse, Robert 1801, "On the Necessary Truth of Certain Conclusions Obtained by Means of Imaginary Quantities," *Philosophical Transactions of the Royal Society of London* 91.

Woodward, James 2005, *Making Things Happen* (Oxford: Oxford University Press).

Woodward, James 2021, *Causation with a Human Face* (Oxford: Oxford University Press).

Woodward, James and Wilson, Mark 2019, "Counterfactuals in the Real World" in N. Fillion, R. M. Corless, and I. S. Kotsireas, eds., *Algorithms and Complexity in Mathematics, Epistemology, and Science* (New York: Springer).

Worrall, John 1989, "Structural Realism: The Best of Both Worlds?," *Dialectica* 43(1–2).

Index